SOLAR POWER TECHNOLOGY

DEVELOPMENTS AND APPLICATIONS

RENEWABLE ENERGY: RESEARCH, DEVELOPMENT AND POLICIES

SOLAR POWER TECHNOLOGY

DEVELOPMENTS AND APPLICATIONS

ANTONIO COLMENAR-SANTOS
ENRIQUE ROSALES-ASENSIO
AND
DAVID BORGE-DIEZ
EDITORS

nova
science publishers
New York

Copyright © 2019 by Nova Science Publishers, Inc.

NOTICE TO THE READER

Library of Congress Cataloging-in-Publication Data

ISBN: 978-1-53614-204-4

Published by Nova Science Publishers, Inc. † New York

CONTENTS

PREFACE

This book constitutes the refereed proceedings of the 2018 International Conference on Solar Power Technology: Developments and Applications, which was held on 28th May 2018. 2018 International Conference on Solar Power Technology: Developments and Applications intends to provide an international forum for the discussion of the latest high-quality research results in all areas related to Solar Power Technology, its developments and applications. The editors believe that readers will find following proceedings interesting and useful for their own research work.

This book contains the Proceedings of the 2018 International Conference on Solar Power Technology: Developments and Applications held online (https://enriquerosales.wixsite.com/virtualconferences), on 28th May, 2018. It covers significant recent developments in the field of Solar Power Technology, its developments and applications from an applicable perspective.

ADVISORY BOARD:

Organizing Committee Chair:

Enrique Rosales Asensio, PhD

Departamento de Física, Universidad de La Laguna, La Laguna, Spain
Email: erosalea@ull.edu.es

PROGRAM COMMITTEE CHAIRS:
Enrique González Cabrera, PhD
Departamento de Ingeniería Química y Tecnología Farmacéutica, Universidad de La Laguna, La Laguna, Spain
Email: eglezc@ull.edu.es
Antonio Colmenar Santos, PhD
Departamento de Ingeniería Eléctrica, Electrónica, Control, Telemática y Química Aplicada a la Ingeniería,
Universidad Nacional de Educación a Distancia, Madrid, Spain
Email: acolmenar@ieec.uned.es
David Borge Diez, PhD
Departamento de Ingeniería Eléctrica y de Sistemas y Automática, Escuela Técnica Superior de Ingenieros de Minas de León,
León, Spain
Email: dbord@unileon.es

SCIENTIFIC COMMITTEE:
Manuel Castro-Gil, Ph.D., Universidad Nacional de Educación a Distancia, Madrid, Spain
Clara M. Pérez-Molina, Ph.D., Universidad Nacional de Educación a Distancia, Madrid, Spain
Francisco Mur-Pérez, Ph.D., Universidad Nacional de Educación a Distancia, Madrid, Spain
Carlos Ignacio Cuviella Suarez, ROCA, Barcelona, Spain
José María Pecharromán Lázaro, ENDESA, Palma de Mallorca, Spain
Elio San Cristobal Ruiz, PhD, Universidad Nacional de Educación a Distancia, Madrid, Spain
Pedro Miguel Ortega Cabezas, PSA, Madrid, Spain

Rosario Gil Ortego, PhD, Universidad Nacional de Educación a
  Distancia, Madrid, Spain
Salvador Ruiz Romero, ENDESA, Barcelona, Spain
Jorge Blanes Peiró, PhD, Universidad de León, León, Spain
Rosario Gil Ortego, Ph.D., Universidad Nacional de Educación a
  Distancia, Madrid, Spain

May 2018

Editors

Antonio Colmenar-Santos
Departamento de Ingeniería Eléctrica, Electrónica, Control, Telemática
y Química Aplicada
Universidad Nacional de Educación a Distancia, Madrid, Spain

Enrique Rosales-Asensio
Departamento de Física
Universidad de La Laguna, La Laguna, Spain

David Borge-Diez
Departamento de Ingeniería Eléctrica y de Sistemas y Automática
Universidad de León, León, Spain

In: Solar Power Technology   ISBN: 978-1-53614-204-4
Editors: A. Colmenar-Santos et al. © 2019 Nova Science Publishers, Inc.

Chapter 1

GRID-CONNECTED PHOTOVOLTAIC FACILITIES AND SELF-SUFFICIENCY

Severo Campíñez-Romero [*] *and Jorge-Juan Blanes-Peiró*
[1]Departamento de Ingeniería Eléctrica, Electrónica, Control,
Telemática y Química Aplicada a la Ingeniería, Universidad
Nacional de Educación a Distancia (UNED), Madrid, Spain
[2]Universidad de León, León, Spain

ABSTRACT

Spain exhibits a high level of energy dependence and has significant solar energy resources. These two facts have given rise to the prominence that renewable energy, particularly solar photovoltaic technology, has enjoyed in recent years, supported by a favorable regulatory framework. Currently, the Spanish government is providing new ways in energy policy to enhance and accelerate the development of low-power photovoltaic generation facilities for self-consumption by introducing energy policies for feed-in payments of surplus electricity. Such facilities are an example of distributed electrical generation with important benefits

[*] Corresponding Author Email: s.campinez.romero@gmail.com.

for the environment and the rest of the electrical system because, when properly managed, they can help improve the system's stability and reduce overall losses. By analyzing household demand and solar photovoltaic energy resources, the profitability of such facilities is considered in this article, taking into account the technical and economic impact of storage systems and proposing models for feed-in payments of surplus electricity, in an attempt to assess whether this method of electricity generation versus the method of conventionally supplied power from a grid at a regulated tariff can rival each other economically, in terms of parity.

Keywords: photovoltaic energy policy, Self-sufficiency household, Storage

NOMENCLATURE

Solar Energy:

G Global irradiance on an optimally inclined plane (W/m^2).

H_{opt} Irradiation on an optimally inclined plane (kWh/m^2).

Photovoltaic (PV) Installation Characteristics:

P_{peak} Peak power of the PV system (kW).

P_{peak}^0 Value of the peak power of an installed PV system that generates surplus electricity capable of being exported to the grid.

SC_{MAX} Maximum storage capacity (% of daily average electricity consumption).

Energy:

E_{gen} Net electricity produced for the PV system (kWh).

E_{load} Household electricity consumption (kWh).

E_{stored} Electricity in the storage system (kWh).

E_{imp} Imported electricity consumed from the grid (kWh).

E_{exp} Exported electricity fed into the grid (kWh).

E_{net} Net electricity exchanged with the grid (kWh).

Costs of Construction, Operation and Maintenance:

C_{inst} PV system construction cost (€/kW).

$C_{storage}$ Storage system cost (€/kWh).

$C_{O\&M}$ PV system operation and maintenance cost (€/kW installed).

C_{INS} Insurance cost (% of C_{inst}).

C_{REP} Electricity market representation cost (€/kWh).

C_{wPV} Electricity consumption cost without the PV system (€).

C_{PV} Electricity consumption cost with the PV system (€).
Tariffs and incomes:

V_{TUR} Present value for the energy term without hourly discrimination of the Last Resource Tariff (€/kWh).

Δ_{TUR} Foreseen annual increase of the Last Resource Tariff (%).

V_{PP} Average final price for the total Spanish demand in the wholesale electricity market (€/kWh).

Δ_{PP} Foreseen annual increase for the average final price for the total Spanish demand in the wholesale electricity market (%).

k_{PP} Increase coefficient for the average final price for the total Spanish demand in the wholesale electricity market.

I Incomes from the sale of electricity exported to the grid (€).

S	Savings in electricity consumption cost achieved by installing the PV system (€).

Financial Terms:

IRR	Internal rate of return of investment over a period of 25 years (%).
PB	Payback period of investment (years).
NPV	Net present value for a period of 25 years (€).
k	Discount rate for NPV calculation.
z	Ordinal indicating the number of years the PV system has been in service. Used in the NPV and IRR calculation.
T_{DEP}	Depreciation period of the PV system (years).
VAT	Value added tax (%).
TAX	Incomes taxes (%).
RPI	Retail price index (%).
CF_z	Cash flow for year z.
CFA_z	Cumulative cash flow for year z.

Note: Superscripts Indicate the Period under Consideration:

ANY^H:	Hourly.
ANY^D	Daily.
ANY^M	Monthly.
ANY^Y	Yearly.

INTRODUCTION

Spain is a country with significant energy dependence; in fact, Spain had an energy dependence level close to 75% in 2010, which is well above the average for EU27 countries of approximately 55% (IDAE. Ministerio de Industria, Turismo y Comercio. Gobierno de España, 2011) for the same period. Reducing this dependence has been one of the main reasons for the strong boost that electricity generation from renewable sources has received from the Spanish government, with a comprehensive development energy policy that was implemented after Royal Decree 2818, became effective in 1998 (Ministerio de Industria y Energía. Gobierno de España, 1998).

Moreover, Spain has significant solar energy resources; it is an EU27 country with relatively high levels of solar radiation. Unlike other types of renewable energy, this resource is the main feature of this article since it is widely available almost everywhere.

In 2009, household electrical consumption was more than 73 million MWh for nearly 24.2 million Spanish consumers; these figures represent 29% of the total consumption and over 85% of the total electricity supply contracts (Ministerio de Industria, Comercio y Turismo. Gobierno de España., 2012).

The important contribution to the total power consumption makes this sector, undoubtedly, a good target for introducing solutions aimed at increasing the use of renewable energy, because every initiative will have an important impact. It also represents a large-sized potential market, which could be translated into a cost reduction associated with an economy of scale.

The current Spanish legal framework provides two options for using electricity from solar energy: stand-alone and grid-connected systems.

In the case of grid-connected systems, this framework, which establishes the remuneration mechanisms as well, began with the enforcement of Royal Decree 2818/1998 (Ministerio de Industria y Energía. Gobierno de España, 1998) and has been updated for the subsequent Royal Decree 436/2004 (Ministerio de Economía. Gobierno de España, 2004), the Royal Decree 661/2007 (Ministerio de Industria, Turismo y Comercio. Gobierno de España, 2007) and later the Royal Decree 1578/2008 (Ministerio de Industria, Turismo y Comercio.

Gobierno de España, 2008) in addition to other complementary legislation. In all cases, the regulatory framework established in this legislation has been organized around the mechanism known as a "feed-in tariff," whit an operation based on guaranteeing grid access and payment for fed electricity above the price established in the electricity market. This cost overrun is funding by the regulated electricity tariff and is, therefore, divided between conventional electricity producers and consumers. As a result of the prioritized system entry of electricity from renewable sources, the resulting price in the wholesale electricity market is reduced.

As a consequence of this legislative framework, the cumulative installed power of grid-connected photovoltaic systems has increased significantly, exceeding the targets set in the Renewable Energy Plan 2005 – 2010, reaching 3,787 GW of installed capacity in 2010 (IDAE. Ministerio de Industria, Turismo y Comercio. Gobierno de España, 2011). However, this large growth was restrained by the enforcement of Royal Decree 1578/2008 and the establishment of power quotas on one side and a gradual reduction of the feed-in tariff (Ciarreta, et al., 2011) on the other. Lately the Royal Law Decree 1/2012 (Jefatura del Estado. Gobierno de España, 2012) has suspended the economic incentives for new electricity production facilities from renewable sources including photovoltaic ones.

In this scenario, new energy policy and regulatory systems will be required in order to assure the growth of implementation of renewable energies in the Spanish energetic mix (Cossent, et al., 2011). Currently, the Ministry of Industry, Tourism and Commerce of the Government of Spain is processing a Royal Decree draft to regulate grid access for low power production installations (Ministerio de Industria, Turismo y Comercio. Gobierno de España., 2012). Such draft establishes mechanisms to facilitate the connection of this type of renewable facility to grids and provides the implementation of a procedure for invoicing and settling the net balance between the electricity produced and consumed. Furthermore, that draft already has the mandatory report of the National Energy Commission (Comisión Nacional de Energía. Gobierno de España, 2011).

On the other hand, from the viewpoint of consumers, Royal Decree 485/2009 (Ministerio de Industria, Turismo y Comercio. Gobierno de España, 2009) enforced the regulatory framework for the establishment of the Last Resource Tariff (LRT, hereafter), which is defined as the price that the last resource retailers can charge for supplying electricity to consumers who have recourse to this law. The enforcement of this new tariff system began July 1, 2009. Currently, there are about 21 million consumers benefiting from the LRT (Ministerio de Industria, Comercio y Turismo. Gobierno de España., 2012) (Comisión Nacional de Energía. Gobierno de España, 2011).

This paper aims to find models for the remuneration of energy generated by small photovoltaic systems, which are mainly designed to support household electrical consumption. These models should provide attractive profitability for users, enhance the investments required and have a positive impact on the whole electric system, due to both the cost of remuneration for the surplus electricity, as well as the provision of ancillary services for grid stability that result from the integration of low-power distributed generation facilities in the distribution electrical grid (Clastres, 2011).

Below, in chapter two, we carry out an estimate of household energy consumption in Spain; in chapter three, we estimate the energy generated by a photovoltaic system located in a representative site. In chapter four, two models for exploiting solar photovoltaic energy will be proposed with an analysis of their operation and establishing the preliminary data to calculate, in chapter five, based on new remuneration frameworks, the profitability of each model and their sensitivity to the most important variables. Finally, in chapter six, we provide a comparison of the models and present the conclusions of the study.

HOUSEHOLD CONSUMPTION DESCRIPTION

In 2007 the electricity consumption of an average Spanish household was 3,992 kWh (Ministerio de Medio Ambiente y Medio Rural y Marino. Gobierno de España, 2012). There are no disaggregated data regarding the evolution of household consumption since then, therefore, this value will be used in this paper as an estimate of the current household demand.

In order to model the daily and yearly variation of household consumption, the results of the INDEL project obtained between 1981 and 1998 (Red Eléctrica de España, S.A., 1998) and published by Red Eléctrica de España, S.A. have been used, assuming that the distribution of actual consumption has not changed substantially since the samples were taken.

The results for the monthly variation of household consumption are shown in Figure 1.

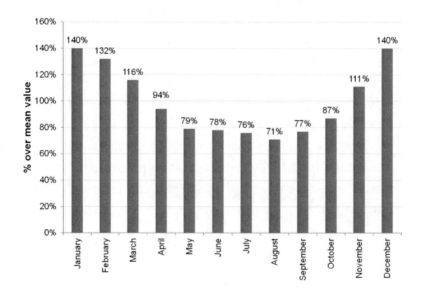

Figure 1. Monthly variation of household consumption. Source: Project INDEL – REE.

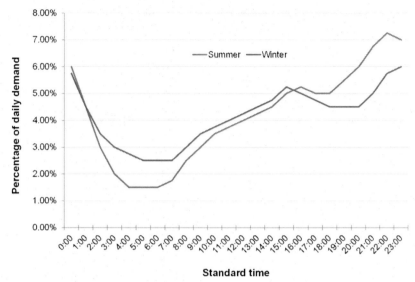

Source: Project INDEL – REE.

Figure 2. Variation of "winter – summer" daily load curves.

A comparison of daily summer and winter load curves is shown in Figure 2.

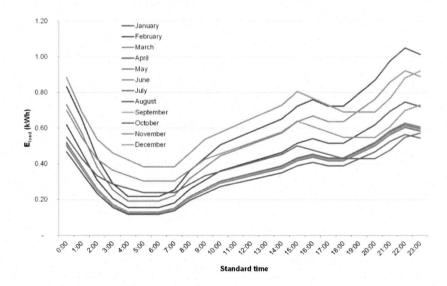

Figure 3. Hourly electricity consumption of an average household.

Table 1. Hourly electricity consumption of an average household

E_{load} (kWh)

Standard Time	January	February	March	April	May	June	July	August	September	October	November	December
0:00	0.88	0.83	0.73	0.62	0.52	0.51	0.50	0.47	0.51	0.55	0.70	0.88
1:00	0.69	0.65	0.57	0.46	0.39	0.38	0.37	0.35	0.38	0.43	0.55	0.69
2:00	0.54	0.43	0.38	0.31	0.26	0.26	0.25	0.23	0.25	0.33	0.43	0.54
3:00	0.46	0.29	0.25	0.21	0.17	0.17	0.17	0.16	0.17	0.29	0.36	0.46
4:00	0.42	0.22	0.19	0.15	0.13	0.13	0.12	0.12	0.13	0.26	0.33	0.42
5:00	0.38	0.22	0.19	0.15	0.13	0.13	0.12	0.12	0.13	0.24	0.30	0.38
6:00	0.38	0.22	0.19	0.15	0.13	0.13	0.12	0.12	0.13	0.24	0.30	0.38
7:00	0.38	0.25	0.22	0.18	0.15	0.15	0.15	0.14	0.15	0.24	0.30	0.38
8:00	0.46	0.36	0.32	0.26	0.22	0.21	0.21	0.19	0.21	0.29	0.36	0.46
9:00	0.54	0.43	0.38	0.31	0.26	0.26	0.25	0.23	0.25	0.33	0.43	0.54
10:00	0.58	0.51	0.44	0.36	0.30	0.30	0.29	0.27	0.30	0.36	0.46	0.58
11:00	0.61	0.54	0.48	0.39	0.32	0.32	0.31	0.29	0.32	0.38	0.49	0.61
12:00	0.65	0.58	0.51	0.41	0.35	0.34	0.33	0.31	0.34	0.41	0.52	0.65
13:00	0.69	0.61	0.54	0.44	0.37	0.36	0.35	0.33	0.36	0.43	0.55	0.69
14:00	0.73	0.65	0.57	0.46	0.39	0.38	0.37	0.35	0.38	0.45	0.58	0.73
15:00	0.81	0.72	0.64	0.52	0.43	0.43	0.42	0.39	0.42	0.50	0.64	0.81
16:00	0.77	0.76	0.67	0.54	0.45	0.45	0.44	0.41	0.44	0.48	0.61	0.77
17:00	0.73	0.72	0.64	0.52	0.43	0.43	0.42	0.39	0.42	0.45	0.58	0.73
18:00	0.69	0.72	0.64	0.52	0.43	0.43	0.42	0.39	0.42	0.43	0.55	0.69
19:00	0.69	0.80	0.70	0.57	0.48	0.47	0.46	0.43	0.46	0.43	0.55	0.69
20:00	0.69	0.87	0.76	0.62	0.52	0.51	0.50	0.47	0.51	0.43	0.55	0.69
21:00	0.77	0.98	0.86	0.70	0.58	0.58	0.56	0.53	0.57	0.48	0.61	0.77
22:00	0.88	1.05	0.92	0.75	0.63	0.62	0.60	0.56	0.61	0.55	0.70	0.88
23:00	0.92	1.01	0.89	0.72	0.61	0.60	0.58	0.54	0.59	0.57	0.73	0.92
Total	15.34	14.43	12.68	10.30	8.66	8.55	8.33	7.78	8.44	9.53	12.16	15.34

From these source data, the hourly distribution of electricity consumption can be obtained for each month. The results are shown graphically in Figure 3 and are detailed in Table 1.

ESTIMATE OF SOLAR PHOTOVOLTAIC ENERGY RESOURCE

Any area of Spanish territory on the Iberian Peninsula has high levels of solar irradiance, specifically, between approximately 1,300 and 2,100 kWh/m^2 annually, reaching up to 2,500 kWh/m^2 annually in the Canary Islands. For the purpose of this article, it is sufficient to assess the photovoltaic energy resources in a location with a mean irradiance, therefore, the city of Madrid was chosen as the PV facility location, due to its central position in the Iberian Peninsula. The Photovoltaic Geographical Information System (PVGIS, hereafter) (European Commission, Joint Research Centre, 2012) was used to evaluate the solar photovoltaic energy resources. This tool provides, among other data, an estimate of the daily electricity produced by the photovoltaic system, taking into account the environmental factors of the selected site. In this case, the input data for the PVGIS are as follows:

- Location: 40°25'0" North, 3°42'1" West, Elevation: 672 m a.s.l.
- Solar radiation database used: PVGIS-CMSAF
- Nominal power of the PV system: 1.0 kW (crystalline silicon)
- Estimated losses due to temperature: 10.3% (using local ambient temperature)
- Estimated loss due to angular reflectance effects: 2.4%
- Other losses (cables, inverters, etc.): 15.0%
- Combined PV system losses: 25.6%

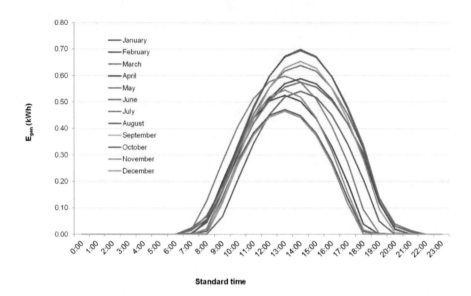

Figure 4. Hourly variation of electricity production for a PV system with 1 kW of installed power. Source: PVGIS and self-elaboration.

In this study, it is essential to know the hourly distribution of electricity produced by the PV system. This hourly distribution is not directly available from the PVGIS tool; however, the hourly distribution of the irradiance received at the site is provided. The raw data are treated to obtain the temporal variation of electricity produced from the temporal distribution of irradiance. The result, for 1 kW of installed power in the PV system, is shown graphically in Figure 4 and numerically in Table 2.

PROPOSED MODELS UNDER STUDY

As shown in Figures 5 and 6, the production (for 1 kW of installed PV power) and consumption do not follow the same pattern; thus the PV system will need support from the grid when it cannot meet consumption demands, but also should be able to feed electricity into the grid during periods in which the production exceeds the demand.

Table 2. Hourly variation of electricity production for a PV system with 1 kW of installed power

E_{gen} (kWh)

Standard Time	January	February	March	April	May	June	July	August	September	October	November	December
0:00	0.00	0.00	0.00	0.00	0.00	0.00	0.00	0.00	0.00	0.00	0.00	0.00
1:00	0.00	0.00	0.00	0.00	0.00	0.00	0.00	0.00	0.00	0.00	0.00	0.00
2:00	0.00	0.00	0.00	0.00	0.00	0.00	0.00	0.00	0.00	0.00	0.00	0.00
3:00	0.00	0.00	0.00	0.00	0.00	0.00	0.00	0.00	0.00	0.00	0.00	0.00
4:00	0.00	0.00	0.00	0.00	0.00	0.00	0.00	0.00	0.00	0.00	0.00	0.00
5:00	0.00	0.00	0.00	0.00	0.00	0.00	0.00	0.00	0.00	0.00	0.00	0.00
6:00	0.00	0.00	0.00	0.00	0.00	0.00	0.00	0.00	0.00	0.00	0.00	0.00
7:00	0.00	0.00	0.01	0.00	0.02	0.02	0.02	0.00	0.00	0.00	0.00	0.00
8:00	0.01	0.05	0.13	0.05	0.07	0.07	0.06	0.04	0.02	0.00	0.02	0.00
9:00	0.14	0.20	0.28	0.16	0.18	0.19	0.18	0.17	0.13	0.07	0.16	0.12
10:00	0.28	0.33	0.41	0.30	0.31	0.33	0.34	0.33	0.29	0.21	0.32	0.27
11:00	0.38	0.44	0.51	0.42	0.42	0.46	0.48	0.48	0.44	0.34	0.44	0.38
12:00	0.45	0.50	0.58	0.51	0.51	0.55	0.60	0.60	0.55	0.45	0.52	0.44
13:00	0.47	0.53	0.60	0.57	0.56	0.62	0.67	0.67	0.63	0.52	0.55	0.47
14:00	0.45	0.50	0.58	0.59	0.58	0.64	0.69	0.70	0.65	0.54	0.52	0.44
15:00	0.38	0.44	0.51	0.57	0.56	0.62	0.67	0.67	0.63	0.52	0.44	0.37
16:00	0.28	0.33	0.41	0.51	0.50	0.55	0.60	0.60	0.55	0.45	0.32	0.26
17:00	0.14	0.20	0.27	0.42	0.42	0.45	0.48	0.47	0.43	0.34	0.16	0.12
18:00	0.01	0.04	0.09	0.30	0.31	0.33	0.34	0.32	0.28	0.21	0.02	0.00
19:00	0.00	0.00	0.00	0.12	0.13	0.14	0.14	0.12	0.10	0.05	0.00	0.00
20:00	0.00	0.00	0.00	0.02	0.04	0.04	0.03	0.02	0.01	0.00	0.00	0.00
21:00	0.00	0.00	0.00	0.00	0.01	0.01	0.01	0.00	0.00	0.00	0.00	0.00
22:00	0.00	0.00	0.00	0.00	0.00	0.00	0.00	0.00	0.00	0.00	0.00	0.00
23:00	0.00	0.00	0.00	0.00	0.00	0.00	0.00	0.00	0.00	0.00	0.00	0.00
Total	2.99	3.56	4.38	4.54	4.61	5.01	5.30	5.19	4.71	3.70	3.46	2.87

Source: PVGIS and self-elaboration.

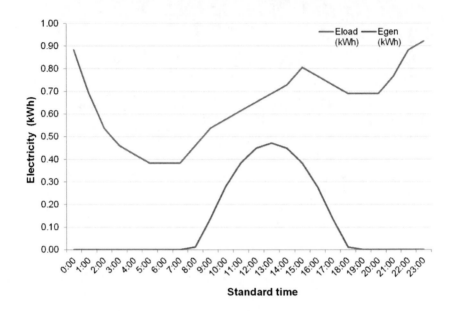

Figure 5. Comparison of PV production and household consumption for January.

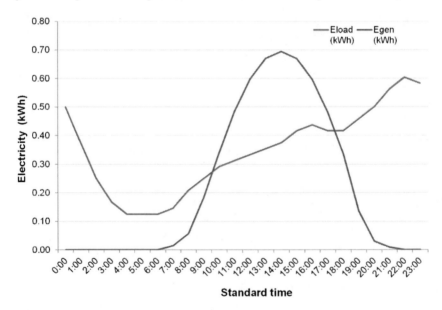

Figure 6. Comparison of PV production and household consumption for July, exhibiting periods with surplus.

This issue can be solved in two ways: either a system with the capacity to offset the electricity fed into grid using the energy supplied by the grid or a system capable of storing surplus electricity for later use. In both cases, there may be surplus electricity that can be used for the rest of the electrical system; consequently, legal mechanisms are needed to remunerate this surplus energy. In this sense, users may recover their investment in two ways: through savings due to self-sufficiency or through income from the surplus energy fed into the grid.

It is essential in this paper that the price to be paid for surplus electricity involves a minimum cost overrun for the whole electrical system; the average final price for the total Spanish demand in the wholesale electricity market V_{PP} will be used as reference.

The compensation for the net electricity balance between the electricity produced and consumed by a facility included in the draft of the Royal Decree proposed by the Ministry of Industry, Tourism and Trade of Spain (Ministerio de Industria, Turismo y Comercio. Gobierno de España., 2012) constitutes a possibility of obtaining profitability that will be discussed in this paper. Furthermore, we will also consider another model based on storing surplus energy to be used during deficit periods.

In short, an analysis of the internal rate of return (IRR, hereafter) and payback period (PB, hereafter) of the investment will be carried out for the following two models:

- Model A: Energy offset and remuneration of net electricity fed into the grid according to the LRT.
- Model B: PV system with stored electricity and remuneration at the average final price for the total Spanish demand in the electricity market.

IRR AND PB CALCULATION

Model A: Energy offset and remuneration of net electricity fed into the grid according to the LRT.

The main scheme for model A is shown in Figure 7.

Electricity produced by the PV system E_{gen} is used to support household consumption E_{load}. Due to the lack of an electricity storage mechanism, this process is instantaneous; as a result, during periods in which the PV system is not capable of meeting the electricity demanded by the household load, additional energy from the grid, E_{imp}, will be required. However, when there is a surplus, it can be fed into the grid; this is represented by E_{exp}. That is:

$$\text{If } E_{gen} < E_{load} \rightarrow E_{imp} > 0 \tag{1}$$

$$\text{If } E_{gen} > E_{load} \rightarrow E_{exp} > 0 \tag{2}$$

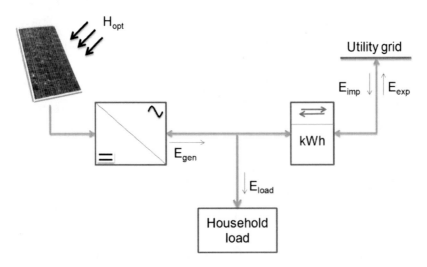

Figure 7. Model A. Main scheme.

The flow of electricity exchanged with the grid will be registered separately in both directions by a bidirectional counter[1]. Later, in the period considered, the net electricity exchanged with the grid will be calculated. Because of the seasonal characteristics of photovoltaic solar energy, an annual period is considered; thus, the net electricity is:

$$E_{net}^y = E_{imp}^y - E_{exp}^y \tag{3}$$

Therefore, E_{net}^y will be positive if, over the whole period considered, it has been necessary to consume electricity from the grid along with the energy generated by the PV system in order to meet the consumption demand, and it will be negative if, during that period, the PV system has been able to meet all consumption demands while generating surplus. The user will have two ways to pay off the installation:

During the periods in which the PV system is not able to meet all consumption alone, the net energy is:

$$E_{net}^y > 0 \tag{4}$$

The income is considered as savings from the consumed energy provided by the PV system. The electricity saved will be valued at the LRT, as this would be the price at which user would have acquired it, meaning that the amount of savings would be:

$$S^y = C_{wPV}^y - C_{PV}^y = \left(E_{load}^y - E_{net}^y\right) \times V_{TUR}^y \times \left(1 + \frac{VAT}{100}\right) \tag{5}$$

For the period in which the PV system is able to meet all consumption demands while generating surplus, the net energy will be:

$$E_{net}^y < 0 \tag{6}$$

In this case, income comes from selling electricity fed to the grid. Because the quantity of net energy is a result of compensation over an annual period, this income is considered as a remuneration value for the surplus energy at the average annual value of V_{PP}^y; therefore, the amount of income would be:

$$I^y = -E_{net}^y \times V_{PP}^y \tag{7}$$

The outcome in each case will depend on the installed PV power as well as on the variations in consumption and generation estimations, meaning that the installations must be designed with an installed capacity to cover at least the electricity requirements, taking into account these fluctuations. Nearly 75% of consumers connect to a low-voltage Spanish electricity grid, corresponding to a contracted power between 1 and 10 kW (Ministerio de Industria, Comercio y Turismo. Gobierno de España., 2012); thus, the current facilities are technically ready for a photovoltaic system with the same installed power, and a large percentage will be designed for a high degree of electrification, which corresponds to an installed capacity of at least 9,200 W (Ministerio de Industria, Turismo y Comercio. Gobierno de España, 2002).

Figure 8 shows the evolution of the income I^y and savings S^y as a function of installed PV power, assuming that both the energy generation and consumption are at the estimated values. Previously, a balance was achieved between the hourly generation and consumption, while calculating the surplus or deficit and integrating over the period considered which, in this case, is one year.

Figure 8 shows a sharp inflection point corresponding to a PV installation capable of meeting household consumption needs and producing surplus electricity to be fed to the grid. This point, called P_{peak}^0, can be calculated based on the household demand and the solar photovoltaic energy resources in the selected location as follows:

$$P_{peak}^{0} = \frac{E_{load}}{E_{gen\left(P_{peak}=1\,kW\right)}} \times 1\,kW = \frac{3{,}992}{1{,}530} \times 1kW = 2.6\,kW \tag{8}$$

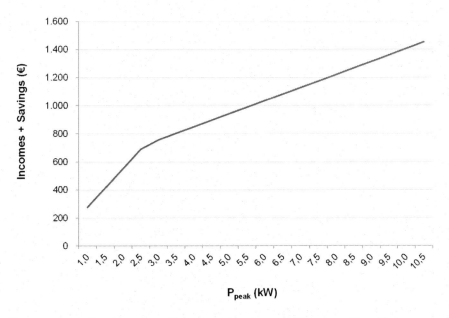

Figure 8. Model A. Evolution of income and savings as a function of installed PV power.

Table 3. Cash flow calculation method

+	Income from surplus electricity remuneration
+	Savings due to demands met by the PV system instead of the grid
-	Operating expenses
=	Gross operating margin
-	Depreciation
=	Earnings before taxes
-	Taxes
=	Earnings after taxes
+	Depreciation
=	Yearly cash flow (CF_z)
+	Cumulative previous year cash flow
=	Cumulative present year cash flow (CFA_z)

The power installation design point should be sufficiently above this value, P_{peak}^0, to absorb the intrinsic variations in solar photovoltaic energy resources and consumption.

Before obtaining the IRR and PB, the cash flows of the investment are calculated following the method described in Table 3.

Here, we take into account the following assumptions:

- The investment is made promptly at the beginning of the period analyzed.
- Incomes are updated yearly with an estimated value for Δ_{PP}.
- Savings are updated yearly with an estimated value for Δ_{TUR}.
- Expenses are updated yearly with an estimated value for RPI.
- External financing is not considered.

The net present value (NPV, hereafter) of the investment is calculated from each yearly cash flow, discounting back to its present value at the discount rate k, that is:

$$NPV = \sum_{z=1}^{n} \frac{CF_z}{(1+k)^z} - I_o \qquad (9)$$

where:

$$I_o = C_{inst} \times P_{peak} \qquad (10)$$

The internal rate of return (IRR, hereafter) is defined as the interest rate k at which the NPV is zero, that is:

$$0 = \sum_{z=1}^{n} \frac{CF_z}{(1+IRR)^z} - I_o \qquad (11)$$

The investment is recovered in the year that the cumulative cash flow exceeds the initial investment. Thus, the payback period (PB, hereafter) is defined as the period required to recover this initial investment through cash flows generated by the installation:

$$PB = z \text{ such } CFA_z \geq 0 \qquad (12)$$

The values given in Table 4 were used to calculate the IRR and PB. The results obtained by calculating the IRR and PB as a function of the installed power are shown in Figures 9 and 10.

Table 4. Model A. Data used for the IRR and PB calculations

Name	Value	Comment
Costs of construction, operation and maintenance:		
C_{inst}	4,000 €/kW	Based on values from installers consulted over the
$C_{O\&M}$	20 €/kW	last six months. Self – elaboration.
C_{INS}	0.5 %	Based on values from consulted insurance companies.
C_{REP}	0.005 €/kWh	According to the Sixth Transitory Provision of Royal Decree 661/2007 (Ministerio de Industria, Turismo y Comercio. Gobierno de España, 2007).
Tariff and incomes:		
V_{TUR}	0.152559 € /kWh	According to the Resolution of December 30, 2011 given by the General Directorate of Energetic Policy and Mines (Ministerio de Industria, Turismo y Comercio. Gobierno de España, 2011).
Δ_{TUR}	3 %	Estimate. Self–elaboration.
V_{PP}	0.0606 /kWh	Corresponding to the mean value between May, 2011 and April, 2012. Source: OMIE (Operador del Mercado Ibérico de Energía Polo Español, 2012).
Δ_{PP}	3 %	Estimate. Self–elaboration.
Financial:		
T_{DEP}	10 *years*	
VAT	18 %	
TAX	18 %	
RPI	3 %	Estimate. Self–elaboration.

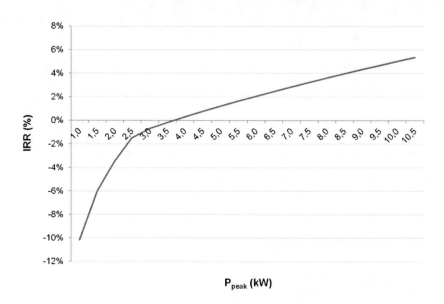

Figure 9. Model A. Evolution of the IRR as a function of installed power.

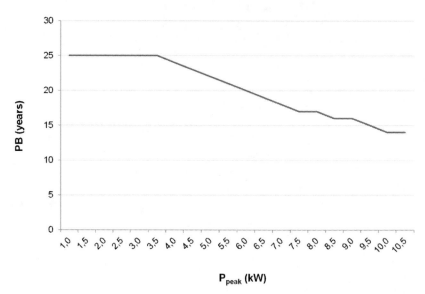

Figure 10. Model A. Evolution of the PB as a function of installed power.

As shown in Figure 9, an IRR over 5% annually could be reached for an installed power of 10 kW. This can be considered as a current market value for financial products with a similar rescue period and could become an alternative to the capital investment required for PV installation. An example is the result of the last auction of 30-year bonds carried out by the Public Treasury of the Government of Spain on May 19, 2011, which resulted in 6.002% of its weighted average rate (Tesoro Público. Ministerio de Economía y Hacienda del Gobierno de España, 2012).

Model B: PV system with stored electricity and remuneration at the average final price for the total Spanish demand in the electricity market

Figure 11 shows the main scheme for model B.

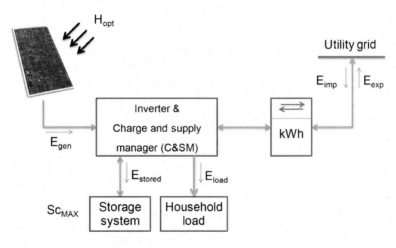

Figure 11. Model B. Main scheme.

In this case the system includes:

- An energy storage subsystem with a maximum capacity SC_{MAX} calculated as a percentage of the daily average electricity consumption. In stand-alone PV installations these systems are typically built with lead-acid batteries and a regulator to manage the charging and discharging cycles. However, there are other possibilities that could be assessed before choosing the storage system (Baker, 2008) (Ibrahim, et al., 2008) (Zahedi, 2011) (Solomon, et al., 2012).

- Equipment to manage both the charging-discharging cycles of the storage system and electricity exchange with the grid.

The operation, shown in Figure 12, is as follows. The electricity produced by the PV system E_{gen} is instantly used to meet the consumption E_{load}. This process can produce two results:

A deficit, that is:

$$E_{gen} < E_{load} \qquad (13)$$

To fill this gap, the charge and supply manager (C&SM, hereafter) may use part of the electricity in the storage system or import energy from the grid. In order to maximize the life of the storage system selected, the maximum depth of discharge has been established in the 20% of the maximum capacity SC_{MAX}. Taking into account this limitation, if the electricity needed to meet the consumption can be obtained from the storage system, the C&SM may allocate part of the stored energy to the load.

$$\text{If } E_{stored} > \left(E_{gen} - E_{load}\right) \text{ then } \left(E_{gen} + E_{stored}\right) \rightarrow E_{load} \qquad (14)$$

In contrast, if the storage system does not have enough energy, the deficit will be filled from the grid.

$$\text{If } E_{stored} < \left(E_{gen} - E_{load}\right) \text{ then } \left(E_{gen} + E_{imp}\right) \rightarrow E_{load} \qquad (15)$$

A surplus, that is:

$$E_{gen} > E_{load} \qquad (16)$$

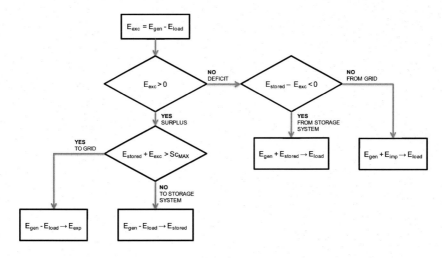

Figure 12. Model B. Operating logic.

In this case, the CS&M must decide where to allocate surplus electricity, giving preference to the storage system versus feeding into the grid.

$$\text{If } E_{stored} + \left(E_{gen} - E_{load}\right) < SC_{MAX} \text{ then } \left(E_{gen} - E_{load}\right) \rightarrow E_{stored} \tag{17}$$

$$\text{If } E_{stored} + \left(E_{gen} - E_{load}\right) > SC_{MAX} \text{ then } \left(E_{gen} - E_{load}\right) \rightarrow E_{exc} \tag{18}$$

Again, the flow of electricity exchanged with the grid will be registered separately in both directions by a bidirectional counter. Later, in the period considered, the net electricity exchanged with the grid will be calculated. In this case, a monthly period is considered because it is used for invoices from distribution companies to consumers.

Electricity imported from the grid will be valued at the LTR. Savings occur from the part of the consumption that was met by the PV system. These savings will be valued at the LTR as well, because this would have been the consumer acquisition price.

$$S^m = CE^m_{wPV} - C^m_{PV} = (E^m_{load} - E^m_{imp}) \times V^m_{TUR} \times \left(1 + \frac{VAT}{100}\right)$$

$$(19)$$

In this case, the amount of electricity fed into the grid is not the result of compensation between the imported and exported electricity, since both are logged at least hourly and could be remunerated at the hourly price for the total Spanish demand in the wholesale electricity market V^h_{PP}; therefore, the income is:

$$I^h = E^h_{exp} \times V^h_{PP} \qquad (20)$$

Once more, the installation must be designed with an installed capacity that can cover at least the electricity requirements, taking weighted fluctuations into account. Furthermore, the effect of the storage system capacity must be properly considered in calculating the income.

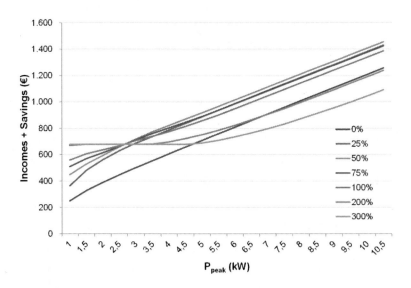

Figure 13. Model B. Evolution of incomes and savings as a function of installed power and storage capacity.

Figure 13 shows the evolution of income I^y and savings S^y as a function of installed PV power, assuming that both energy generation and consumption are at the estimated values. According to model A, it was necessary to reach a balance between the hourly generation and consumption in calculating the surplus or deficit, but in model B, we consider the storage capabilities, giving priority to the use of storage electricity against energy imported from the grid. An annual integration has again been made to obtain the results.

It can clearly be seen that even for a low installed power a high value of SC_{MAX} may be the best choice. From a certain point on, a storage limit of 50% provides the most revenue. This boundary corresponds to P^0_{peak}, the installed power value at which the PV system fully satisfies the energy needs and begins to generate surplus to be fed into the grid.

The method used for obtaining the IRR and PB for model B is the same as that used for model A, but considers the cost of the storage system, so the value of the investment cost is now:

$$I_o = C_{inst} \times P_{peak} + C_{storage} \times SC_{max} \times 3{,}992 \text{ kWh} \qquad (21)$$

The values given in Table 5 were used to calculate the IRR and PB in this case.

The results for the IRR and the PB obtained from model B calculations are given in Figures 14 and 15. They show that, despite the incomes and savings are greater for a 50% of storage capacity installed, a value of 25% offers better profitability figures due to its lower cost. That is the reason why this value will be used in the rest of calculations.

As for the model A, the results indicate that an IRR over 5% annually could be reached by an installed power over 10 kW.

Table 5. Model B. Data used for the IRR and PB calculations

Name	Value	Comment
Photovoltaic (PV) installation characteristics:		
SC_{MAX}	25 *and* 50 %	As shown in Figure 13, the storage capacity value that maximizes the income and savings is 50%. Nevertheless, taking into account that a 25% of storage capacity offers very similar values but a lower cost, IRR and PB will be calculated for a value of 25% too, in order to assure if the better results for incomes and savings mean a better profitability.
Costs of construction, operation and maintenance:		
C_{inst}	4,000 €/kW	Based on values from installers consulted over the last six
$C_{O\&M}$	20 €/kW	months. Self – elaboration.
$C_{storage}$	500 €/kWh	This cost corresponds to a solution based on lead-acid electrical storage batteries. In the calculations, it was assumed that 50% of the storage capacity is renewed every 5 years, beginning with the initial service date, and this price is updated as the annual increase in the consumer price index RPI
C_{INS}	0.5 %	Based on values from consulted insurance companies.
C_{REP}	0.005 € /kWh	According to the Sixth Transitory Provision of Royal Decree 661/2007 (Ministerio de Industria, Turismo y Comercio. Gobierno de España, 2007).
Tariff and incomes:		
V_{TUR}	0.152559 € /kWh	According to the Resolution of December 30, 2011 given by the General Directorate of Energetic Policy and Mines (Ministerio de Industria, Turismo y Comercio. Gobierno de España, 2011).
Δ_{TUR}	3 %	Estimate. Self – elaboration.
V_{PP}	0.0606 /kWh	Corresponding to the mean value between May, 2011 and April, 2012 Source: OMIE (Operador del Mercado Ibérico de Energía Polo Español, 2012).
Δ_{PP}	3 %	Estimate. Self–elaboration.
Financial:		
T_{DEP}	10 *years*	
VAT	18 %	
TAX	18 %	
RPI	3 %	Estimate. Self–elaboration.

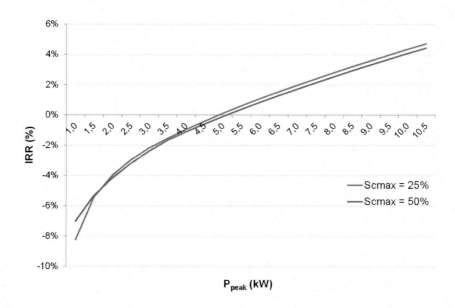

Figure 14. Model B. Evolution of the IRR as a function of installed power for a storage capacity of 25% and 50%.

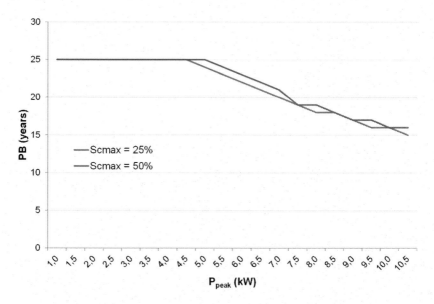

Figure 15. Model B. Evolution of the PB as a function of installed power for a storage capacity of 25% and 50%.

COMPARISON OF RESULTS FOR BOTH MODELS

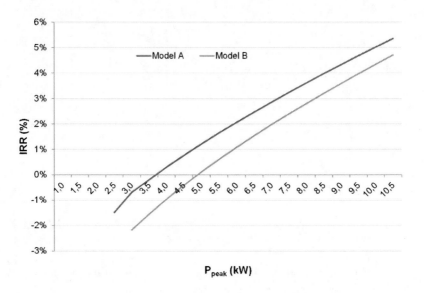

Figure 16. Comparison of evolution of the IRR as a function of installed power for both models.

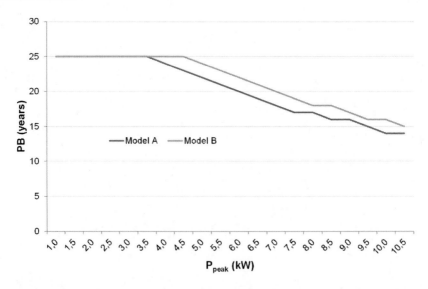

Figure 17. Comparison of evolution of the PB as a function of installed power for both models.

Under the assumptions indicated in the preceding sections, and as shown in Figures 16 and 17, the model A provide better results, approximately 1% of IRR and around two years for PB.

SENSITIVITY ANALYSIS OF THE IRR AND PB IN RELATION TO THE REMUNERATION OF SURPLUS ELECTRICITY

One way to improve the IRR and PB results obtained is to increase the remuneration of surplus energy, which was valued at the final price of the domestic demand for electricity in the wholesale market V_{PP}^y.

In order to consider variations in the income calculations, a factor that increases the price, called k_{PP}, has been included. This way, the modified equations for the income are as follows:

For model A:

$$I^y = -E_{net}^y \times k_{PP} \times V_{PP}^y \tag{22}$$

For model B:

$$I^h = E_{exp}^h \times k_{PP} \times V_{PP}^h \tag{23}$$

The results are shown in Figures 18 to 21.

There is a clear improvement in results, achieving internal rates of return greater than 10% and payback periods of less than 10 years. Besides, for a required profitability, the installed power of the facility decreases with increasing incomes.

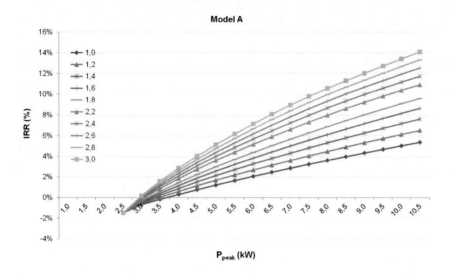

Figure 18. Model A. Evolution of the IRR as a function of installed power and k_{pp}.

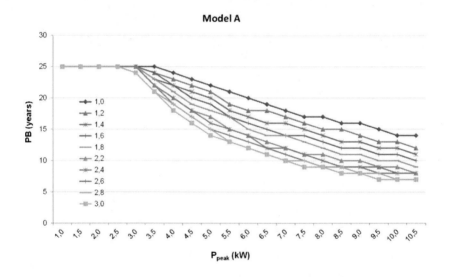

Figure 19. Model A. Evolution of the PB as a function of installed power and k_{pp}.

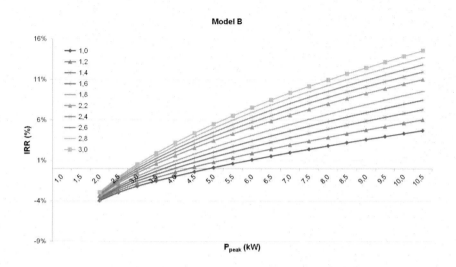

Figure 20. Model B. Evolution of the IRR as a function of installed power and k_{pp}.

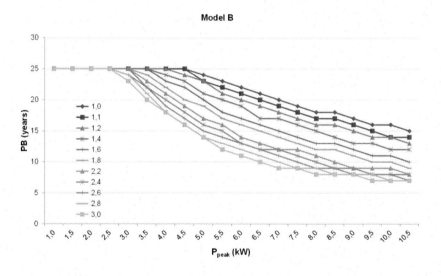

Figure 21. Model B. Evolution of the PB as a function of installed power and k_{pp}.

It should be noted that an increase in k_{PP} corresponds to an increase in the cost of the whole electric system for purchasing the surplus energy. However, we must also consider that the feed-in tariff for the year 2012 (Ministerio de Industria, Turismo y Comercio. Gobierno de

España, 2011) for type I.1 facilities[2], as defined in RD 1578/2008 (Ministerio de Industria, Turismo y Comercio. Gobierno de España, 2008), which are studied in this paper, is 0.330866 €/kWh. This represents an increase of 546% over the annual average final price of the domestic demand for electricity in the wholesale market considered (0.0606 €/kWh). Accordingly, an appropriate remuneration would be equivalent to setting $k_{PP} = 5.46$, which is well above the maximum value of the range considered in the study, $k_{PP} = 3$.

SENSITIVITY ANALYSIS OF THE IRR AND PB IN RELATION TO INSTALLATION COST VARIATIONS

Another way to improve the IRR and PB obtained is to reduce investment costs. Overall, it should be considered that the breadth of the potential market may involve important cost savings, due to an economy of scale.

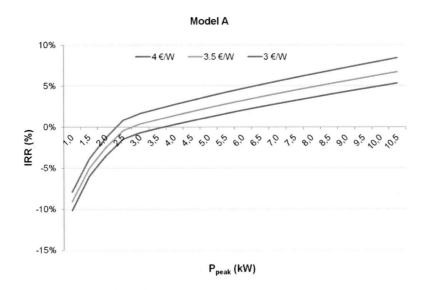

Figure 22. Model A. Sensitivity of the IRR to the investment cost.

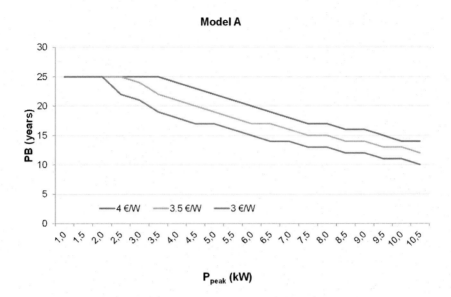

Figure 23. Model A. Sensitivity of the PB to the investment cost.

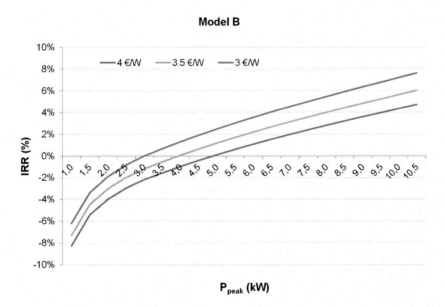

Figure 24. Model B. Sensitivity of the IRR to the investment cost.

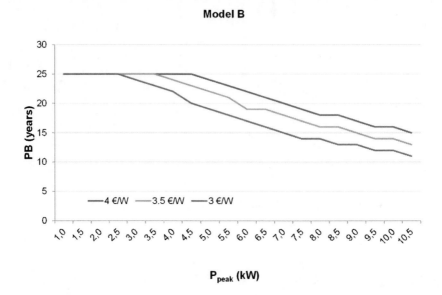

Figure 25. Model B. Sensitivity of the PB to the investment cost.

Among all of the possible costs, the dominant one is the cost of the PV system, ignoring the storage system cost $C_{storage}$. The results for the IRR and PB as a function of the installed power and installation cost for $k_{PP} = 1$ while keeping the other variables fixed are shown in Figures 22 to 25.

As expected, a reduction in investment costs results in an increase in the IRR and a reduction in the PB.

CONCLUSION

In the light of the above in this work, we can say that new energy policy and regulatory frameworks should be available in the near future in order to allow the large-scale integration of distributed generation facilities. Moreover, those frameworks must include retributive mechanisms for the surplus energy.

Household electricity self-sufficiency achieved by using solar photovoltaic energy can be profitable in economic terms, if the sale of surplus electricity is allowed at prices well below the current feed-in tariff for grid-connected PV installations. This represents important savings for the whole electric system, since large amounts of electricity can be used from renewable sources at a much lower cost than its remuneration from the current feed-in tariff.

In relation to the proposed models for the remuneration of surplus energy generated, the model A gives a slightly higher profitability but is more expensive for the whole electric system, because by offsetting exported and imported electricity, part of the surplus is remunerated at the LRT. On the other hand, model B does not require as much electricity from the grid, since it uses stored energy during periods of deficit.

Although the model based on the storage of surplus electricity gives lower profitability ratios, additional income could be achieved if stored energy is properly delivered to the grid during hours when the price of electricity in the wholesale market is higher, with variations of up to 150% in value.

Furthermore, systems with energy storage can contribute to electric grid stability with ancillary services (Clastres, et al., 2010) (Zamora & K. Srivastava, 2010) (Wade, et al., 2010) by providing reactive energy and aiding in voltage regulation in the distribution. Those activities should be remunerated, but the corresponding income was not considered in this paper because of the current lack of an appropriate legal framework.

It should also be taken into consideration that PV facilities are an example of distributed generation with important benefits for the environment (Funsho Akorede, et al., 2010). In addition, they would provide an important reduction in power losses in electricity transport and distribution; it is difficult to make an exact estimate of this reduction, but considering that power losses in transport and distribution on the Iberian Peninsula were 4.7% of the electricity available in 2009 (Ministerio de Industria, Comercio y Turismo. Gobierno de España., 2012), we could say that, for approximately every kWh produced by the photovoltaic system to support a household, the whole system would reduce its power losses by similar percentage.

Households in cities are mostly gathered in blocks of flats that clearly would not allow for the installation of enough power to meet all consumption demands, but, even so, this study shows that with sufficient increases in storage capacity, acceptable profitability could be achieved. In fact, profitability may even be higher than the 25% considered for PV systems with an installed power greater than P_{peak}^0.

REFERENCES

[1]   Baker, J., 2008. New technology and possible advances in energy storage. *Energy Policy,* Issue 36, pp. 4368-4373.
[2]   Ciarreta, A., Gutiérrez-Hita, C. & Nasirov, S., 2011. Renewable energy sources in the Spanish electricity market. Instruments and effects. *Renewable and Sustainable Energy Reviews,* Issue 15, pp. 2510-2519.
[3]   Clastres, C., 2011. Smart grids: Another steps towards competition, energy security and climate change objectives. *Energy Policy,* Issue 39, pp. 5399-5408.
[4]   Clastres, C., Pham, T. H. & F. Wurtz, S. B., 2010. Ancillary services and optimal household energy management with photovoltaic production. *Energy,* Issue 35, pp. 55-64.
[5]   Comisión Nacional de Energía. *El consumo eléctrico en el mercado peninsular en el año 2010* [*Electricity consumption in the peninsular market in 2010*] Madrid: s.n; 2011.
[6]   Comisión Nacional de Energía. Gobierno de España, 2011. *Informe 17/2011 de la CNE sobre el proyecto de Real Decreto por el que se regula la conexión a red de instalaciones de producción*

de energía eléctrica de pequeña potencia (aprobado por el Consejo en su sesión de 2 de junio de 2011), [CNE Report 17/2011 on the draft Royal Decree regulating the connection to the network of small power plants (approved by the Council at its meeting on 2 June 2011)]; Madrid: s.n.

[7]   Cossent, R., Gómez, T. & Olmos, L., 2011. Large-scale integration of renewable and distributed generation of electricity in Spain: Current situation and future needs. *Energy Policy,* Issue 39, pp. 8078-8087.

[8]   de la Hoz, J. et al., 2010. Promotion of grid-connected photovoltaic systems in Spain: Performance analysis of the period 1998-2008. *Renewable and Sustainable Energy Reviews,* Issue 14, pp. 2547-2563.

[9]   European Commission, Joint Research Centre, 2012. *Photovoltaic Geographical Information System (PVGIS).* [Online] Available at: http://re.jrc.ec.europa.eu/pvgis/[Accessed 2012].

[10]   Funsho Akorede, M., Hizam, H. & Poruesmaeil, E., 2010. Distribuited energy resources and benefits to the environment. *Renewable and Sustainable Energy Reviews,* Issue 14, pp. 724-734.

[11]   Ibrahim, H., Ilinca, A. & Perron, J., 2008. Energy storage systems. Characteristics and comparisons. *Renewable & Sustainable Energy Reviews,* Issue 12, pp. 1221-1250.

[12]   IDAE. Ministerio de Industria, Turismo y Comercio. Gobierno de España, 2011. *Plan de energías renovables de España (PER) [Plan of renewable energies of Spain] 2011 - 2020,* s.l.: s.n.

[13]   Jefatura del Estado. Gobierno de España, 2012. *Real Decreto-Ley 1/2012 de 27 de enero, por el que se procede a la suspensión de los procedimientos de preasignación de retribución y a la supresión de los incentivos económicos para nuevas instalaciones de producción de energía eléctrica [Royal Decree-Law 1/2012 of 27 January, which proceeds to the suspension of the pre-allocation of remuneration procedures and to the elimination of*

economic incentives for new electric power production facilities];
2012.

[14] Ministerio de Economía. Gobierno de España, 2004. *Real Decreto*
436/2004, de 12 de marzo, por el que se establece la metodología
para la actualización y sistematización del régimen jurídico y
económico de la actividad de producción de energía eléctrica en
régimen especial [*Royal Decree 436/2004, of 12 March,*
establishing the methodology for updating and systematizing the
legal and economic regime of the activity of production of electric
energy under special regime].

[15] Ministerio de Industria y Energía. Gobierno de España, 1998. *Real*
Decreto 2818/1998, de 23 de diciembre, sobre producción de
energía eléctrica por instalaciones abastecidas por recursos o
fuentes de energía renovables, residuos y cogeneración [*Royal*
Decree 2818/1998, of 23 December, on the production of
electricity by facilities supplied by renewable resources or sources
of energy, waste and cogeneration].

[16] Ministerio de Industria, Comercio y Turismo. Gobierno de
España., 2012. *Estadísticas eléctricas anuales. Estadísticas y*
balances energéticos [*Annual electrical statistics. Energy*
statistics and balances]. [Online] Available at: http://www.
minetur.gob.es/energia/balances/Publicaciones/ElectricasAnuales/
Paginas/ElectricasAnuales.aspx [Accessed 2012].

[17] Ministerio de Industria, Turismo y Comercio. Gobierno de
España., 2012. *Propuesta de Legislación - Energía Eléctrica -*
Energía - Mº de Industria, Turismo y Comercio [*Legislation*
Proposal - Electric Power - Energy - Ministry of Industry,
Tourism and Trade] [Online] Available at: http://www.
minetur.gob.es/energia/electricidad/Legislacion/Documents/20110
411_Propuesta_RD_conexiones.pdf[Accessed 2012].

[18] Ministerio de Industria, Turismo y Comercio. Gobierno de
España, 2002. *Reglamento Electrotécnico de Baja Tensión e*
Instrucciones Técnicas Complementarias [*Low Voltage*

Electrotechnical Regulation and Complementary Technical Instructions] s.l.:s.n.

[19] Ministerio de Industria, Turismo y Comercio. Gobierno de España, 2007. *Real Decreto 661/2007 de 25 de mayo por el que se regula la actividad de producción de energía eléctrica en régimen especial* [*Royal Decree 661/2007 of 25 May regulating the activity of production of electricity in special regime*] s.l.:BOE.

[20] Ministerio de Industria, Turismo y Comercio. Gobierno de España, 2008. *Real Decreto 1578/2008, de 26 de septiembre, de retribución de la actividad de producción de energía eléctrica mediante tecnología solar fotovoltaica para instalaciones posteriores a la fecha límite de mantenimiento de la retribución del RD 661/2007* [*Royal Decree 1578/2008, of 26 September, on the remuneration of the activity of production of electric energy by means of photovoltaic solar technology for installations after the deadline for maintenance of the remuneration of RD 661/2007*] s.l.:BOE.

[21] Ministerio de Industria, Turismo y Comercio. Gobierno de España, 2009. *Real Decreto 485/2009, de 3 de abril, por el que se regula la puesta en marcha del suministro de último recurso en el sector de la energía eléctrica* [*Royal Decree 485/2009, of 3 April, regulating the start-up of the supply of last resort in the electric energy sector*]. s.l.:BOE.

[22] Ministerio de Industria, Turismo y Comercio. Gobierno de España, 2011. *Orden IET/3586/2011, de 30 de diciembre, por la que se establecen los peajes de acceso a partir de 1 de enero de 2012 y las tarifas y primas de las instalaciones del régimen especial* [*Order IET / 3586/2011, of December 30, establishing the access tolls from January 1, 2012 and the tariffs and premiums of the facilities of the special regime*] s.l.:BOE.

[23] Ministerio de Industria, Turismo y Comercio. Gobierno de España, 2011. *Resolución de 30 de diciembre de 2011, de la DGPEyM, por la que se establece el coste de producción de*

energía eléctrica y las tarifas de último recurso a aplicar entre el 23 y el 31 de diciembre de 2011 y el primer trimestre de 2012 [*Resolution of 30 December 2011, DGPEyM, which establishes the cost of production of electricity and the tariffs of last resort to be applied between 23 and 31 December 2011 and the first quarter of 2012*].

[24] Ministerio de Medio Ambiente y Medio Rural y Marino. Gobierno de España, 2012. *Banco público de indicadores ambientales del Ministerio de Medio Ambiente, y Medio Rural y Marino* [*Public Bank of Environmental Indicators of the Ministry of the Environment, and Rural and Marine Environment*] [Online] Available at: http://www.magrama.gob.es/es/calidad-y-evaluacion-ambiental/temas/informacion-ambiental-indicadores-ambientales/banco-publico-de-indicadores-ambientales-bpia-/default.aspx#para3 [Accessed 2012].

[25] Operador del Mercado Ibérico de Energía Polo Español, 2012. *Operador del Mercado Ibérico de Energía Polo Español* [*Operator of the Iberian Energy Market Spanish Polo*]. [Online] Available at: http://www.omie.es/files/flash/ResultadosMercado.swf [Accessed 2012].

[26] Red Eléctrica de España, S.A., 1998. *Atlas de la demanda eléctrica en España* [*Atlas of electricity demand in Spain*]. Madrid: Red Eléctrica de España, S.A.

[27] Solomon, A., Faiman, D. & Meron, G., 2012. Appropriate storage for high-penetration grid-connected photovoltaic plants. *Energy Policy,* Issue 40, pp. 335-344.

[28] Tesoro Público. Ministerio de Economía y Hacienda del Gobierno de España, 2012, *Tesoro Público.* [*Ministry of Economy and Treasury of the Government of Spain, 2012, Public Treasury*] [Online] Available at: http://www.tesoro.es/sp/subastas/resultados/o_30a_11_05_19.asp [Accessed 2012].

[29] Wade, N., Taylor, P., Lang, P. & Jones, P., 2010. Evaluating the benfits of an electrical energy storage system in a future smart grid. *Energy Policy,* Issue 38, pp. 7180-7188.

[30] Zahedi, A., 2011. Maximizing solar PV energy penetration using energy storage technology. *Renewable and Sustainable Energy Reviews,* Issue 15, pp. 866-870.

[31] Zamora, R. & K. Srivastava, A., 2010. Controls for microgrids with storage: Review, challenges, and research needs. *Renewable and Sustainable Energy Reviews,* Issue 14, pp. 2009-2018.

In: Solar Power Technology   ISBN: 978-1-53614-204-4
Editors: A. Colmenar-Santos et al. © 2019 Nova Science Publishers, Inc.

Chapter 2

MODELING OF MULTI-MEGAWATT PV PLANTS

Tomás Guinduláin-Argandoña[*]
and Enrique-Luis Molina-Ibáñez

Departamento de Ingeniería Eléctrica, Electrónica, Control,
Telemática y Química Aplicada a la Ingeniería, Universidad
Nacional de Educación a Distancia (UNED), Madrid, Spain

ABSTRACT

This chapter develops an integrated model of multi megawatt PV plant with HVDC (High Voltage Direct Current) or HVAC (High Voltage Alternating Current) network, using the specific software of power electronics PSIM. This model has been developed by functional blocks, including the photovoltaic field itself, the pertinent conversion units for the integration of each network as well as the network type for production. The models allow to obtain transmissions loss for any combination of the three variables on which they depend; network length (km), temperature (ºC) and irradiance (W/m2). To verify the validity of

---

[*] Corresponding Author Email: el.molina-ibanez@gmail.com.

the model and demonstrate the distribution advantages of HVDC -even for relatively low-photovoltaic power plants in comparison to the common applications currently used in HVDC networks-, a case study has been used which has led to the conclusion that the use of HVDC networks may be convenient for this type of power generation plants.

Keywords: simulation of modeling of multi-megawatt photovoltaic plants with high voltage direct current grid integration

ACRONYMS

MMWPV	Multi-megawatt Photovoltaic
HVAC	High Voltage Alternating Current
HVDC	High Voltage Direct Current
HVDCGR	High Voltage Direct Current Ground Return
HVDCMR	High Voltage Direct Current Metallic Return
NOTC	Nominal Operating Cell Temperature
STC	Standard Test Conditions

INTRODUCTION

This research re-sparks the well-known *War of Currents* of the 19th century that had Edison, a proponent of DC and Tesla as the standard bearer of the AC [1]. From the current perspective, it is a challenge to discern a single winner; it is a long-established fact that such technological disagreement will result in a technical tie between both forms of electric transport and distribution, given the current technological momentum experienced by HVDC (*High Voltage Direct Current*) technology.

Such momentum arises from its advantages, highly contrasted and derived from the absence of the capacitive and inductive phenomena in AC networks. Such capabilities can be summarized as follows [2]:

- Great transport efficiency. DC can carry large amounts of electrical power over long distances.
- Flexibility as it permits asynchronous interconnections.
- Suitable for submarine links.

These great benefits have propelled the proliferation of HVDC networks and consequently gaining ground to alternating current by transporting and distributing more megawatts year after year. Proof of this is the number of complete HVDC projects, in progress or about to be completed. A few of these projects are emblematic. Two of these examples include the HVDC link DC voltage to ± 500 kV of 2004 in *Three Gorges (Guangdong, China)*, capable of transmitting 3000 MW over 940 km [3] or, the HVDC interconnection link between France and Spain of October 2015, which transmits ± 320 kV and 1400 MW of transport capacity [4].

Much like its advantages, there is a plethora of literature devoted to its disadvantages, the following are worth noting [2]:

- Higher cost of DC converters.
- Higher harmonic generation

The intricate control employed in multi-terminal operations. The majority of the links are point-to-point.

But perhaps the main technical hurdle that has had to overcome HVDC technology has been the lack of a suitable switch [3]; when an AC switch is opened, an arc continues driving the current between the contacts to the next zero crossing. Since the DC current does not have this useful pass through the zero value of the current, a different approach is needed, and this has for a long time prevented the development of more complex HVDC network topologies [3].

However, these drawbacks are being addressed and or minimized by new switches capable of opening circuits with high alternating current, such as the hybrid switch developed by ABB. This switch combines semiconductor technology for the rapid interruption of DC with a fast mechanical switch [5]. In addition, the costs of the converter stations are decreasing, thanks to the increase of production in the sale of units and their greater integration in the market.

A direct consequence of HVDC's growing demand is the high expectations set forth by the sector in renewable energy generation based on its high efficiency in transportation and long distance transmission capabilities while incurring minimum transmission loss, giving Renewable Energies a fresh and hopeful boost. This is the case, for example, of the *TuNur project* [6], consisting of a large solar thermal plant that is intended to be built in the Sahara of Tunisia to supply electricity to Europe through a submarine cable of more than 600 km connected to the grid European in the Italian part.

In this instance, it makes sense to use photovoltaic power plants in direct current since PV modules already produce direct current electricity from sunlight. Additional factors encourage a paradigm shift in the distribution of energy generated in these plants. These factors are as follows;

- The increased proliferation of DC powered equipment, such as the electric car, or LED lighting has prompted the review of low-voltage electrical distribution, and in some cases is being considered to generating power straight into DC, developing networks with a greater efficiency and cost reduction [7].
- The birth of a new type of solar PV power stations, of floating type, on areas of water reservoir, such as the one in *Nishihira Pond* of 1,7 MW or 1,2 MW *Higashihara's pond*, both in the city of *Kato, Japan*, in operation since March 2015 [8]. This revolutionary design has increased production thanks to its cooling effect, solving the lack of space while reducing the evaporation of water reserves [9].
- Although the development of these floating photovoltaic fields is focused on reservoirs or the cited ponds, it opens the door for its use in maritime zones, emulating offshore wind farms. In fact, there are already PV modules on the high seas, prototypes and testing techniques for future use [10].
- In both cases, submarine type links would be required, and under these conditions, AC transport is limited to short distance, given the high capacitance of isolated conductors, as already indicated.

Unlike the typical applications of HVDC networks such as large and long-distance energy transports, connections between asynchronous systems and submarine links, this research has a different focus. This chapter explores its use in similar networks such as the large photovoltaic power plants in the range of several megawatts, but with inferior power typically handled by HVDC links.

Therefore, the fundamental objective of this study is the development of a simulation model for a grid connected multi-megawatt photovoltaic plant and its different technologies HVAC and HVDC. Such model can precisely detect energy losses incurred during distribution based on two of three variables that affect transmission, with the fixed network length; temperature, and irradiance.

In this way, it will be possible to have a very useful tool in obtaining the critical distance of the links, in other words, the distance from which is most profitable to distribute in DC instead of AC, and that as a general rule is approximately between 800 and 900 km for aerial networks and 60 and 70 km for underground [11].

Another key aspect is verifying the suitability of small-scale HVDC links to distribute the production of multi-megawatt photovoltaic plants and determine if these plants can benefit from said advantages, which would undoubtedly lead to a greater integration and presence of Solar photovoltaic energy worldwide.

In *Section II*, the various stages of work -that have been covered to achieve the objectives set out in the first section- are discussed along with a technical analysis of the conditions and functional requisites for the integration of the MMWPV plants into HVDC networks. This analysis culminates in a proposal of unit converter plant MMWPV - Network HVDC that will lay the foundation for the rest of the study. Subsequently, two working models of MMWPV plant are developed, including both HVDC and HVAC networks, Figures 2 and 3.

In *Section III*, the models generated are applied to a specific case of MWPV plant, using real technical data of the integrating components and gathered from commercial catalogs. These elements have been previously modeled in a generic way (as already indicated) to obtain through the subsequent simulation the losses resulting from energy transportation according to the variables which define them including—temperature (°C) and irradiance (W/m^2), and considering a specific length. This section concludes with a summary of results, Table 4.

Finally, *section IV* presents the most relevant conclusions based on the results including, the validity of the models, the technical and economic viability of the integration of MWPV into HVDC and the possible applications in both developed and developing countries.

MATERIALS AND METHODS

The simulation models were built upon a well differentiated four phase process:

1. The phase of research and analysis of the PV plant elements including - distribution networks, with emphasis on the current situation and future trends or lines of development, taking as references, besides those indicated, the technical documentation of the leading manufacturers in the transport sector in HVDC, *ABB*, and *Siemens*.
2. Block structured functional design of the required central converting unit (multi-megawatt photovoltaic plant connected to HVDC grid).
3. Modeling the system, using PSIM power electronics software, following a modular strategy; photovoltaic field, central converter unit MMW PV connected to HVDC grid, central converter unit MMW PV- HVAC, HVDC and HVAC distribution networks.
4. A case study to apply the generated models and subsequent simulations to measure transmit losses for any combination of the variables that affect transmission rate including, temperature (°C) and irradiance (W/m^2), with a certain network length.

   These simulations will be tested on three types of networks (HVDC GR, Ground Return, HVDC MR, Metallic Return, and Three-phase HVAC), with aerial or underground lines.

   By following this process, all the possible cases will be covered, establishing the validity of the generated models, in addition while comparing transmit loss among these to showcase the advantages of HVDC transmission for PV plants, which fall under the category of small-scale plants for electric generation.

FUNCTIONAL DESIGN CENTRAL CONVERTING UNIT MMWPV-HVDC NETWORK

To carry out the coupling of PV plants in HVDC networks it will be necessary a DC/HVDC converter capable of coupling the variable production in voltage and DC type and low voltage that occurs in the PV field to the high voltage requirements in continuous and variable current of the HVDC networks.

There are several alternatives to raise the voltage of the DC field, DC type and variable according to the environmental conditions, to the orders of magnitude of the order of kV required in the HVDC links (since the Joule effect losses decrease with the square of the tension). The alternatives are:

- Connect the strings of modules in series or strings. A no-viable option due to the limit of the inverse voltage in the PV modules, around the 1000 V.
- Connect the strings of modules in series or strings after the maximum power point tracker, MPPT or Maximum Power Point Tracker, which is also not viable because in practice it would disable its functionality, subduing all the modules to the maximum power generated by the chain of worst production.
- Another option could be including a boost converter (or booster) after the MPPT block; a structure already offered by a few current PV inverters without a transformer. These converters are primarily used in regulated DC power sources and regenerative DC motor braking. As its name implies, the output voltage is always larger than the input voltage, and it includes a large output filter capacitor to ensure a constant output voltage [12]. This option, however, has two limitations; one in the amplification range, too short for the needed

requirements and the other is low performance compared to DC/AC converters.

• From the standpoint of achieving equal performance to that of the DC/AC converters of the PV plants in AC networks, the best option to achieve the high output voltages of the PV-HVDC unit is, paradoxically, using an AC transformer. The proposed structure is similar to those seen in the current medium voltage photovoltaic plans or MMSS (Multi Megawatt Solar Station), but with an additional rectifier stage.

These platforms are making headway in the last two years favored by the programming and installation of the major photovoltaic systems, mainly in the United States, China and several countries in South America, in the order of several tens of megawatts [13].

Medium-voltage PV platforms power ranges from 300 kW to 2500 kW contain all the necessary elements to simplify the construction of high-power photovoltaic solar energy installations. These vary depending on the manufacturer parameters such as the number of MPPT trackers, the number and type of DC/AC converters, and the characteristics of the low to medium voltage transformer.

By adding a rectifier stage to one of these platforms, the outcome is a structure capable of satisfying the functional requirements, i.e., a high and constant DC tension, which will vary depending on the solar production of each moment. In other words, it is reversed to raise the tension and rectified to reap the benefits of transmission in HVDC.

This structure is a combination of two standalone systems; the medium voltage PV platform on one side and a small-scale AC/HVDC converter station of the order of megawatt with order output voltages of some tens of kV.

Therefore, this PV-HVDC unit must have (see Figure 1):

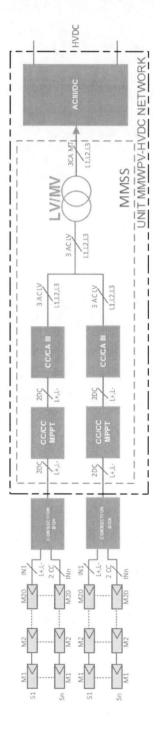

Figure 1. Functional block diagram of the proposed MMWPV-HVDC station (self-developed).

- MPPT (maximum power point tracking) blocks, which are truly DC converters. These trackers are fundamental to increase the efficiency in the utilization of the production of the photovoltaic field. Much has been done in the recent years regarding improvements in control algorithms to improve tracking. One of the future trends may include tracking algorithms based on adaptive control techniques [14].

- Central type DC/AC converter units, given the many modules connected into the MMW PV plant; it is customary to have an inverter for every 500 kW approximately. It is proposed to use the three-phase VSI (voltage source inverter) type with SPWM (pulse width sinusoidal modulation), state-of-the-art control technology based on the utilization of multiple modulations with pulse width variation proportional to the amplitude of a sine wave evaluated at the center of said pulse [15]. Given the power requirements, this VSI inverter must use IGBT (Insulated Gate Bipolar) switches. There will be a constant value sine wave voltage at the output of these inverters along with a variable current which follows the current variations of the PV field.

- A single low to medium voltage transformer to collect the outputs of all the DC/AC converters the platform supports. A single transformer could be included for each inverter, but that would increase the final price. It is important to point out that a few types of medium-power grid-connected inverters include high-frequency transformers to boost performance to which the previous CC/CA bridge works at elevated frequencies.

- A final rectifier stage based on IGBT type devices is recommended, given the need to reach medium voltage tensions, this final stage must be composed of a multi-level converter of three levels, according to the Diode-Clamped topology also known as Neutral-Point-Clamped, with two capacitors on the DC bus. Multilevel inverters are a breakthrough concerning traditional inverters, as these can handle large voltages and powers, generating alternating voltages with fewer harmonics than conventional inverters [16]. The diodes connected to the midpoint are the elements that set the blocking voltages of the switches at a fraction of the DC bus voltage, making these an essential component of this topology. During normal operations, there are two semiconductor switches per phase in a locked state, which allows double-voltage operation of the DC link of a two-level converter using the same elements. As a result, this converter/rectifier will allow working with output voltages in the medium voltage range, making the VSI bridge with IGBT combination the best choice [17].

Once the necessary conversion unit structure is selected, the next step is developing a valid functional model of all the elements that comprise the MMW PV plant, from the PV modules to the distribution network, to procure its integration into a single structure, which is also the final objective of this study.

The specialized power electronics software PSIM V11.0.3 [18] will be used for this purpose and a modeling line will be followed according to specific functional criteria specific to the elements that make up the installation (for example, the PV modules) together with the rules and electrotechnical laws that govern the electrical behavior of the assembly.

Modeling of the PV Field and the MPPT Block

As previously indicated, PSIM is a critical resource for this research and consequently for this modeling as it aids in the implementation of the photovoltaic field. This physical solar model can be found in PSIM's component library "Renewable Energy Module". This module factors the variation of solar radiation and temperature, which are fundamental to the analysis of losses during distribution under different environmental conditions (reference PV field in Figure 2 & Figure 3). PSIM PV model block was used to model our PV system and modelled the MPPT with an efficiency gain. The equivalent circuit that employs the physical model includes the equivalent circuit of a solar cell, along with the equations that characterize it [18].

Nevertheless, it is necessary to supply a large number of parameters. A few of these can be found in the manufacturer's data sheets as shown below;

- N_S, number of serially arranged cells which compose the module.
- V_{OC} (V), open circuit voltage.
- V_{MP} (V), the voltage at maximum power.
- I_{SC} (A), short-circuit current.
- I_{MP} (A), current at maximum power.
- P_{PMP} (W), maximum power, peak power or at the point of maximum power.
- V_{OC} temperature coefficient (%/°C o K).
- I_{SC} temperature coefficient (%/°C o K).
- (dv/di) in V_{OC} refers to the slope (dv/di) in the open circuit voltage V_{OC}, which can be estimated from the V-I characteristic curve of the solar module.

Table 1 displays other parameters, these parameters are not usually provided by the datasheet, so in [19] is presented how to obtain these parameters with the help of the I-V curve provided by the datasheet [20]. *Solar Module (Physical Model)* is another useful feature included in PSIM as it facilitates the definition of the solar module parameters described in section III.B, and display precisely the operation of the chosen module for the application case.

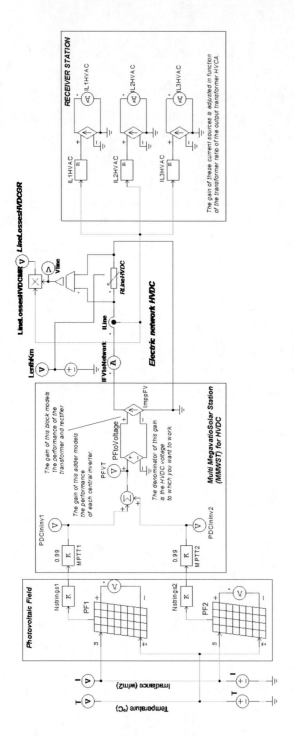

Figure 2. Complete MMW PV plant to HVDC Model (self-developed).

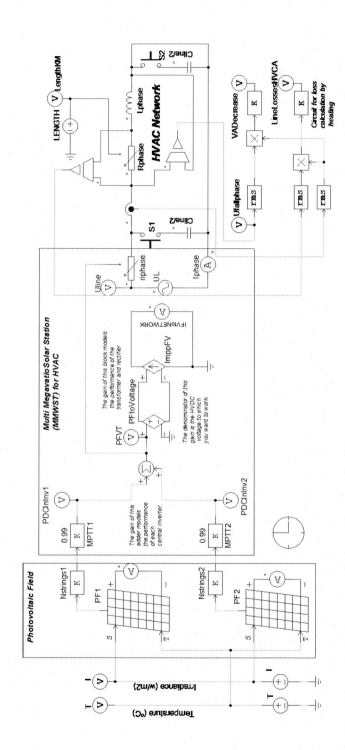

Figure 3. Complete MMW PV plant to HVAC Model (self-developed).

Table 1. PV Module Required Parameters [19]

Parameters to be Defined	
Band Energy Eg	eV
Coefficient Ks	No units
Ideality Factor A	No units
Shunt Resistance Rsh	Ω
Parameters to Be Calculated	
Series Resistance Rs	Ω
Saturation Currnet Is0	A
Short Circuit Current Isc0	A
Temperature Coefficient Ct	A/°C o K

The MPPT block can be implemented following different control strategies, such as [21]:

- Method of disturbance and observation (D&O).
- Incremental conductance method.
- Hill Climbing Technique.
- MPPT method with current measurement.

The physical model of the PV panel that has been used to model the PV field, has 2 inputs, where the temperature and irradiance conditions of the module are set, and three outputs:

- One positive and one negative terminal to deliver the power the module is giving, per temperature and irradiance as well as the connected load.
- The third output is at the top, where maximum power point information is provided about those ambient conditions regardless of the connected load. This output will, therefore, be utilized since it is equipped with the maximum power point tracker.

A proportional block with a gain of 0,99 has been included to take into account the current performance of the MPPT devices, which is around 99%.

PV-HVDC Unit Modeling

At a functional level, the converter-transformer-rectifier unit behaves as a constant-current transducer, causing the HVDC output network to have a high voltage, of the order of kV, and a fixed value, and a current variable that follows the variations produced in the PV field. Therefore, a model has been developed based on the following criteria as indicated in Figure 2:

- The use of an adder is required to sum the power produced by each of the converter units, providing us with the PPVT, total PV power. The value of the converter unit's performance can be included in each of the input gain, which can be estimated generically at 0,95.
- By adding a voltage-to-voltage transducer, it is possible to adapt the working voltage for the rest of the elements of the circuit. The gain of this transducer will be adjusted according to the working voltage of the HVDC network. Therefore, the amount of gain from the first transducer is a DC voltage of equal value to the PV current.

$$I_{MPPPV\ TOTAL} = \frac{PPVT}{V_{HDVC}} \tag{1}$$

- The next step involves another voltage-current transduction, adjusting the gain of this second transducer to the estimated transformer performance.

As a result, the PV-HDVC unit will have all the current produced by the PV field with the corresponding performances and the desired working voltage.

PV-HVAC Unit Modeling

The operation of MMW PV platforms with medium distribution AC system modeling is required to comparing transmit losses between the HVDC or HVAC networks. Such modeling calls for an alternating voltage generation with an effective value which corresponds to the working HVAC voltage and resistance that changes according to the PV output as indicated in Figure 3.

The variation of the voltage's internal resistance based on production is equivalent to injecting more or less current into the HVAC network. The use of the *nonlinear element V = f (i, x)* is necessary to consider the association of PV HVAC system to establish the parameters. In this equation, the PV power generation is the external variable based on environmental conditions.

$$V_{LINE\ HVAC} = f(i,x) = i_{phase} \cdot r_{i\,phase} = i_{PHASE} \cdot \frac{V_{HVDC}^2}{\left(\dfrac{PPVT}{\sqrt{3}}\right)} =$$

$$= \frac{i_{PHASE} \cdot V_{HVDC}^2 \cdot \sqrt{3}}{PPVT} \tag{2}$$

For the parameterization of the system it is necessary to know the voltage trend with respect to the intensity, as shown below;

$$\frac{\partial V_{LINE\ HVAC}}{\partial i} = \frac{V_{HVDC}^2 \cdot \sqrt{3}}{PPVT} \tag{3}$$

HVDC Network Modeling

Before modeling a monopolar link, there is a need to include a resistance that characterizes the ohmic value of the line, which will depend on the type of conductor, the distance between the link and the calculating section. The network model HVDCGR, nonlinear element v = f (i, x) is necessary to include the effect of the variation of the link length to establish the parameters. In this equation, the length is the external variable, which will change during the simulation from zero to the link's maximum distance.

$$V_{LINE\ HVDCGR} = f(i,l) = I \cdot R_{LINEHVDC} = i \cdot \rho \cdot \frac{l}{S} .1000 \tag{4}$$

The multiplication factor of 1000 is included to work in km. For the parameterization of the system it is necessary to know the voltage trend with respect to the intensity, as shown below;

$$\frac{\partial V_{LINE\ HVDCGR}}{\partial i} = \rho \cdot \frac{l}{S} .1000 \tag{5}$$

When modeling a monopolar link with a metallic return, it is only required to include the multiplication factor of 2 in expressions 4, and 5 since the distance to consider is doubled due to the use of two conductors in the link.

Table 2. Input Variables and indicators for the HDVC Model (self-developed)

Name	Function	Unit
Irradiance (Voltage Source)	Irradiance adjustment	W/m^2
μ CC/CA (In block adder)	AC/DC Converter Performance adjustment	No unit
μ MPTT (in proportional block)	MPPT block performance	No unit
μ Transformer (Voltage-controlled current source)	Transformer performance adjustment	No unit
IFVtoNetwork (Current Probe)	PV current to network	A
IL1HVAC (Current Probe)	AC current to HVAC network (phase 1)	A
IL2HVAC (Current Probe)	AC current to HVAC network (phase 2)	A
IL3HVAC (Current Probe)	AC current to HVAC network (phase 3)	A
Irradiance, I (Voltage Probe)	Irradiance indicator	W/m^2
LenthKm (Voltage Probe)	HVDC link length indicator	km
LenthKm (Voltage Probe)	HVDC length adjustment	km
LineLossesHVDCGR (Voltage Probe)	HVDC GR voltage drop	W
LineLossesHVDCMR (Voltage Probe)	HVDC MR voltage drop	W
N° Strings (in proportional block)	PV Field adjustment	No unit
PDCInInv1 (Voltage Probe)	DC power produced by converter 1	W
PDCInInv2 (Voltage Probe)	DC power produced by converter 2	W
PPVT (Voltage Probe)	Power in DC produced by all converters	W
RLineHVDC	HVDC link resistance	Ω
Temperature (Voltage Source)	PV Module operating temperature adjustment	°C
Temperature, T (Voltage Probe)	PV Module operating temperature indicator	°C
VHVDC (Voltage-controlled voltage source)	HVDC link voltage	V
Vline (Voltage Probe)	HVDC link voltage drop	V

Table 3. Input Variables and Indicators for the Hdac Model [self-developed]

Name	Function	Unit
μ CC/CA (In block adder)	AC/DC Converter Performance adjustment	No unit
μ MPTT (in proportional block)	MPPT block performance	No unit
μ Transformer (Voltage-controlled current source)	Transformer performance adjustment	No unit
Cline	HVAC capacitance	μF
IFVtoNetwork (Current Probe)	PV current to network	A
Iphase	HVAC link current (per phase)	A
Irradiance (Voltage Source)	Adjust irradiance level	W/m^2
Irradiance, I (Voltage Probe)	Irradiance indicator	W/m^2
LenthKm (Voltage Probe)	HVAC link length indicator	km
LenthKm (Voltage Source)	HVAC link length adjustment	km
LineLossesHVAC (Voltage Probe)	Loss of power in the HVAC link	W
Lphase	HVAC link inductance (per phase)	mH
N° Strings (in proportional block)	PV Field adjustment	No unit
PDCInInv1 (Voltage Probe)	DC power produced by converter 1	W
PDCInInv2 (Voltage Probe)	DC power produced by converter 2	W
PPVT (Voltage Probe)	Power in DC produced by all converters	W
riphase	HVAC internal resistance (per phase)	Ω
Rphase	HVAC internal resistance (per phase)	Ω
S1, S2	Allows to take into account or not the capacitive	No unit
Temperature (Voltage Source)	PV Module operating temperature adjustment	ºC
Temperature, T (Voltage Probe)	PV Module operating temperature indicator	ºC
VADecrease	HVAC link diminished capacity	VA
VFallPhase (Voltage Probe)	HVAC link diminished capacity	VA
Vline	RMS line voltage in the HVAC link	V

HVAC Network Modeling

The total inductive reactance value of each phase is estimated based on the data provided by the manufacturer for insulated cables, and design of aerial networks with bare conductors. Once this value is known and to determine the value of the required inductance all that is required is to apply the concept of reactance impedance:

$$X_{LPhase} = 2 \cdot \pi \cdot f \cdot L(H) \rightarrow L(H) = \frac{X_{LPhase}}{2 \cdot \pi \cdot 50} \qquad (6)$$

Simulation and Variable Modeling

Once the previously mentioned systems are modeled the full integration yields the following models: MMW PV-HDVC as seen in Figure 2, and MMW PV-HVAC as seen in Figure 3. Tables 2 and 3 specify the meaning of all external variables and measurement points for both models.

RESULTS AND DISCUSSION

Installation Data to Be Modeled

Through modelling it might be able to estimate transmission losses for any combination of the variables that affect transmission including, temperature (°C) and irradiance (W/m^2), with a certain network length. If no measurements were available, the best way to test the veracity of these models is by running simulations on a few actual case studies involving, a 2 MW PV power plant for HVDC and

HVAC connection, a Ground Return HVDCGR and metallic return HVDCMR.

The baseline data includes:

- Types of distribution networks to be analyzed:
  - An 80-km underground distribution network built with insulated aluminum cables. Total transit loss will be calculated in the three modeled networks and different scenarios of irradiance and temperature.
  - An 80-km aerial distribution network built with bare aluminum cables. Total transit loss will be calculated in the three modeled networks and different scenarios of irradiance and temperature.
- The following commercial elements will be considered for the parameterization of the model:
  - *ABB* medium-voltage PV array of 2MW nominal power, model PVS800-IS-2000kW-C with the following specifications for the parameterization purposes, are;
    - Number of MPPT units = 2
    - Number and type of inverter units: 2. Central inverter type. 2 x 1200 kW of maximum input power.
    - Number of inputs per inverter: 12
    - MPPT input voltage range: 600 to 850 V
    - Maximum input voltage: 1100V
  - Commercial PV module model NU-RC290 | 290 W from the manufacturer Sharp [20].
  - Medium-Voltage range commercial conductors, i.e., high-modulated ethylene-propylene insulated cables, called EPROTENAX will be used for the underground network. The technical specifications are extracted from [23].

- Bare aluminum conductors, i.e., LA-XX type, will be used for the aerial network. The technical specifications are extracted from [23].

• Several case studies will be analyzed for each type of network (aerial or underground), and type of distribution technology (HVDCGR, HVDCMR, and HVAC), to establish a sound comparison of transmission losses between them. The cases consist of 4 relevant situations in each type of network according to the following alpha-numerical sequence: A (Aerial Network), U (Underground Network), Temperature (ºC)/ Irradiance (W/m^2).

- U/A Cases -10ºC/1000 (W/m^2), relevant cases because they representing the extreme limit of operation for the calculation of the upper limit of string couplings to the inverter units.

- U/A Cases 25/1000 (W/m^2), relevant cases because they representing the standard conditions of measurement or STC.

- U/A Cases 48/800 (W/m^2), relevant cases because they representing the normal operating temperature conditions of the cell, or NOTC.

- U/A Cases 70/1000 (W/m^2), relevant cases because they representing the extreme limit of operation for the calculation of the lower limit of string couplings to the inverter units.

These cases have been selected because they are very characteristic values in the PV studies in terms of irradiance and temperature, such as the STC and NOTC conditions. The other values are within the operating temperature of the electrical elements and meet the requirements of III.C. The value of -10ºC is due to the fact that in the Shetland Islands, the minimum registered temperature is -8.9ºC according to [24].

These simulations are carried out considering the distance variable of 80 km because it can be considered as an acceptable distance for PV installations on islands, floating or even in developing countries.

Once the general installation data is established, said data would then be parametrized in the generated models to calculate the transmission losses through simulation.

Solar Module Parametrization

The solar module used for this case is the module NU-RC290, manufactured by SHARP, consisting of 60 monocrystalline silicon cells connected in series [20].

It is necessary to use the tool *Solar Module (Physical Model),* to transfer the functional characteristics of the module to the physical panel model. This process consists of the following steps [19].

III.B.1 Datasheet Introduction

The basic electrical parameters I_{MPP}, V_{MPP}, I_{SC}, V_{OC}, P_{PMP}, the number of cells and temperature coefficients of I_{SC}, V_{OC}, are introduced.

The value "dv/di (slope) in V_{OC}" refers to the slope dv/di to the open-circuit voltage Voc, which can be approximated by the V-I curve of the module, obtaining an estimated value of -0,5.

III.B.2 Initial Estimation of EG, A, RSH, Y KS

Typically, these four parameters are not provided in the data sheet and have to be approximated.

According to the considerations of [19], in the case of the NU-RC290 module, we will assume initial values of: Eg = 1,12V, A = 2, Rsh = 2000 Ω, Ks = 0.

III.B.3 Calculating I-V P-V Curves and the Maximum Power Point

With the previous information, the tool calculates the following parameters: *series resistance R_S, short-circuit current $Isc0$,* saturation current *Is0*, and temperature coefficient *Ct*.

III.B.4 Comparison of the Data Sheet and Experimental Data for Different Operating Conditions, and Parameter Adjustment

Since the first parameter approximation is outside its nominal value in the catalog, it is necessary to fine tune the parameters A and (dv/di), generating a more accurate value which represents the selected module for the case study as shown in Figure 4.

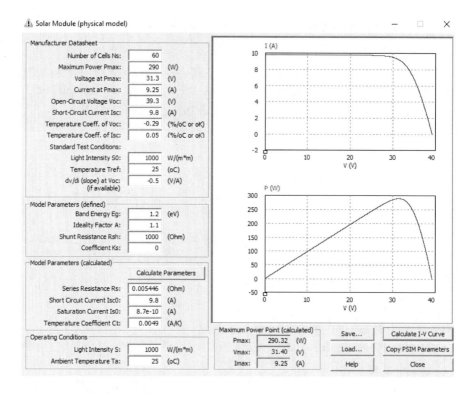

Figure 4. Parameterization of the UN-RC290 module with the Solar Module (Physical Model) tool (self-developed).

A fine-tuned parametrization reveals a closer value to that of the nominal power, which is 290,32W≈290W. These values are then loaded into the two physical panel models in the diagram, depicting the operation of the NU-RC290 module accurately.

Photovoltaic Field Configuration

Given the voltage at the output of the PV panels is a function of temperature, it is necessary to verify the inverter can operate in adverse conditions (-10°C to + 70°C). The optimum coupling between the PV field and the inverter will, therefore, be guaranteed provided the followings conditions are met:

$$U_{minPV}\left(@1000W / m^2 y\ T^a = 70^\circ C\right) \geq U_{MPPT\,min} \tag{7}$$

$$U_{maxPV}\left(@1000W / m^2 y\ T^a = -10^\circ C\right) \leq U_{MPPT\,max} \tag{8}$$

According to the module and inverter unit technical data sheets, the following apply:

- Solar module
  - UOC module = 39,3 V
  - Variation of Voltage with temperature: -0,29%/°C.
- Inverter Unit MPPT data coupling
  - Inverter UMPPT min = 600 V
  - Inverter UMPPT max = 850 V
  - Inverter U max = 1100V

The previously mentioned conditions-presented in 7 and 8- can be verified by serially connecting the modules; however, in this instance and since each inverter unit is equipped with 12 inputs, it is necessary to interconnect these in 17 parallel strings, consisting of 20 modules per string.

$$20 \left(\frac{modules}{string}\right) \times 17 \left(\frac{string}{paralell}\right) \times 12 \left(\frac{inputs}{inverter}\right) = 4080 \left(\frac{modules}{inverter}\right)$$
(9)

Thus, the maximum total power for each inverter unit will be as follows:

$$P_{TOTAL\ INVERTER} = 4080 \left(\frac{modules}{inverter}\right) \times 290 \left(\frac{Wp}{module}\right) = 1\ 183\ 200\ W \approx 1200kW$$
(10)

Before implementing this PV field distribution in the model, the following must occur:

- The 20-module series per strings is configured by multiplying the number of cells of each physical model by 20:

$$20 \left(\frac{modules}{string}\right) \times 60 \left(\frac{cells}{module}\right) = 1200 \left(\frac{cells}{string}\right)$$
(11)

- The parallel association of the 17 module strings series and the connection to each of the 12 inputs is configured by including the proportional block called N_{string} (see Figures 2 and 3) with equal gain is shown below:

$$N_{STRINGS} = 4080(\text{modules/inverter})/20\ (\text{modules/string}) =$$
$$= 204\ (strings\ /\ inverter)$$
(12)

FV-HVDC Converter Unit Settings

Once both PV module parameterization and PV array configuration are complete, the next step involves the parameterization of the FV-HVDC converter. Given the simplicity of the implemented model, only the desired HVDC voltage must be included. Such voltage is included in the transducer gain block power PV to tension (see Figures 2 and 3).

There must be a correlation between the power and the voltage carried. Ideally, it is best to work at the highest allowed voltages, as losses due to the Joule effect decrease with the square of it. Furthermore, it is equally important to consider the range of standard distribution voltages.

A good way to approximate the appropriate voltage value in a distribution is by using the *Still* expression

$$U_{NETWORK}(kV) = 5,5 \cdot \sqrt{\frac{0,62 \cdot l + P}{100}} = 24,8 kV \tag{13}$$

Therefore,

U voltage between phases, in kV

l line length in km

P Average power to be transmitted in kW

The distribution voltage of 20 kV, 20 000 V is a very common distribution voltage. It also matches that of the power to be transmitted. Therefore, the transductor gain power-voltage model should be configured as follows;

$$G_{TRANSDUCTOR\ PPV\ to\ V} = V_{RED\ HVDC} = V_{RMS\ RED\ HVAC} = 20 kV \tag{14}$$

PV-HVAC Converter Unit Settings

The PV-HVAC converter unit settings involve the parametrization of the generator's internal resistance of 20 000V between lines to correlate as already indicated, the PV production into an HVAC network. By applying the expression (2), the nonlinear riphase resistance must be parametrized. The external variable is the generated

PV power along with the already known working voltage of 20kV as shown below:

$$V_{LINE\ HVAC} = f(i,x) = i_{phase} \cdot r_{i\,phase} = i_{PHASE} \cdot \frac{V_{HVDC}^2}{\left(\dfrac{PPVT}{\sqrt{3}}\right)} =$$

$$= \frac{i_{PHASE} \cdot 20\,000^2 \cdot \sqrt{3}}{PPVT} \tag{15}$$

For the parameterization of the system it is necessary to know the voltage trend with respect to the intensity, which is:

$$\frac{\partial V_{LINE\ HVAC}}{\partial i} = \frac{20\,000^2 \cdot \sqrt{3}}{PPVT} \tag{16}$$

Underground Network Parameterization

The total length of the underground network is 80 km, and it consists of 20 kV insulated cables suitable for underground or underwater applications.

The following data is required to model the pertinent circuit for the underground network:

- Calculating conductor sections to obtain the resistance for HVDCGR and HVDCMR respectively.
- Inductive reactance X_L and capacity C equivalent per km and per phase are both necessary to model the three-phase HVAC link along with the resistance value obtained in the previous case.

The calculation of the section requires-given the length of the link and economic criteria to reducing values to acceptable values- the use of higher sections of those technically required based on the criteria of nominal load currents and short circuit current. It is important to remember that the voltage drop criterion has no bearing when calculating the section in a medium voltage network.

- As a starting point, power losses (associated with the heating of the conductors) is around 3% in the HVDCGR case.
- The previously calculated section will be used to compare the losses in the three types of links.

Considering the fundamental equation of transport as shown below [25]:

$$Losses(\%) = K \cdot \frac{l}{S} \cdot \frac{P_1}{U_1^2} \qquad (17)$$

Therefore,

- K = Coefficient which depends on the line type.
- l = Length of the line.
- S = Section of the conductor used.
- P1 = Power measured at the source.
- U1 = Voltage measured at the source.

The coefficient K has the following value:

- DC lines: $K = 2\rho$
- Single-phase AC lines: $K = \dfrac{2\rho}{cos^2\varphi_1}$
- Three-phase AC lines: $K = \dfrac{\rho}{cos^2\varphi_1}$

Its application reveals that in order to limit losses in an aluminum line to 3%, with a resistivity of 1/28 (m/Ω mm^2) at 90°C; therefore, a section of 476 mm^2 is required:

$$S = \frac{1}{28} \cdot \frac{80\,000}{0{,}03} \cdot \frac{2\,000\,000}{20\,000^2} = 476\,mm^2 \tag{18}$$

In this case, a section greater than 500 mm^2 is appropriate to minimize losses at the end of the HVDC GR run (80 km). It is a suitable section based on the total transport length (abnormally high) for the (small-scale) range of power to be distributed, especially when the purpose is to compare both networks, HVDC and HVAC.

The cable specifications reveal the inductance and capacity of the lines. In this case, and according to the specifications obtained in [23], the following information applies:

- The chosen section can handle a much higher current than the one circulating for each case, as clearly indicated by the economic criterion to minimize transmission losses.
- The same applies to the intensity criterion of the short-circuit current.
- The capacity of the cable, according to the manufacturer's data is 0,484 (μF/km), reactance at 0,095 (Ω/km), and in the case of three unipolar cables, the value is of 500 mm^2. These values are acceptable values because of the high capacitance and a low reactance, typical characteristics of insulated wires.

The parameterization of the nonlinear Rline resistance is required in the HVDCGR case to adjust the length during the simulations. Considering a maximum operating temperature of 90°C and aluminum conductor and applying (4)

$$V_{LINE\ HVDCGR} = f(i,l) = I \cdot R_{LINEHVDC} = i \cdot \rho \cdot \frac{l}{S}.1000 = i \cdot \frac{1}{28} \cdot \frac{l}{500}.1000$$

$$(19)$$

For the parameterization of the system it is necessary to know the voltage trend with respect to the intensity, which is

$$\frac{\partial V_{LINE\ HVAC}}{\partial i} = \frac{1}{28} \cdot \frac{l}{500}.1000 \tag{20}$$

For the modeling of the underground HVAC network, the following will be considered:

- Nonlinear and variable resistance with the length of each phase is modeled just like the HVDCGR, parameterization equations (19) and (20) since it is the same section, same conductor and same distance.
- In the case of an underground line with a high capacitance factor per km, the switches S1 and S2 connecting the capacitors at the ends of the configuration to "pi" shall be closed, assigning each a value of:

$$C_{LINE/2} = \frac{0,484 \times 80}{2} = 19,36\mu F \tag{21}$$

- The inductance value of the line shall be, applying (6), and considering a factor of 0,095 (Ω/km) and a length of 80 km:

$$X_{LPhase} = 2 \cdot \pi \cdot f \cdot L(H) \rightarrow L(H) = \frac{0,095 \times 80}{2 \cdot \pi \cdot 50} = 24,19mH \tag{22}$$

Aerial Network Parameterization

The aerial network is 80 km and must use bare 20 kV cables between phases, in addition to the supports and other elements necessary for its layout. The design includes a simple symmetric phase in a triangle. A commercial example of suitable cables for medium voltage aerial distribution is bare aluminum cables type LA-XX. The technical specifications can be extracted from [23].

The commercial section closest to that calculated based on economic criteria is 500 mm2 for underground networks, so the same example will be used again to achieve a more homogeneous comparison. A bare aluminum conductor with a 500 mm^2 section.

From this aluminum section and considering the link's length, upgrading HVDC networks is like the one underground.

The following must be considered when modeling an HVAC aerial network:

- Nonlinear and variable resistance with the length of each phase is modeled as in the case of HVDC GR, parametrization equations (19) and (20) since it is the same section, same conductor and same distance.
- Since it is a short-length aerial line with bare conductors, the effect of the capacitance can be ignored. As a result, switches S1 and S2−connecting the capacitors at the ends based on the "pi" configuration−will open.
- The line's inductance value shall be, applying (6), considering a factor of 0,35 (Ω/km), a common value for aerial lines with a symmetrical in equilateral triangle design [13] and a length of 80 km:

$$X_{LPhase} = 2 \cdot \pi \cdot f \cdot L(H) \rightarrow L(H) = \frac{0,35 \times 80}{2 \cdot \pi \cdot 50} = 89,12 mH \quad (23)$$

Simulation Settings

Once the necessary operations are performed to parameterize all the elements of both models, there is a need to adjust the parameters and begin the simulation.

Settings of the Temporary Parameters of the Simulation

The temporary parameters for the graphical simulation are adjusted through the *Simulation Control tool*. Samples will be taken every 0,001 seconds. The entire simulation involving all variables will last 20 seconds.

Voltage (v) -Irradiance (w/m²) Transduction

The solar modules reach a working voltage level starting at a small little irradiance value, typically from 2 to 8 (mW/cm²). An intermediate starting point 50 (W/m²) will be taken, and therefore the voltage line required to cover the irradiance range, from 50 to 1000 (W/m²), must be set to 20 seconds of simulation, as:

$$u(t)_{\to I(W/m^2)} = 50 + \frac{1000-50}{20} \cdot t = 50 + 47,5 \cdot t \qquad (24)$$

Voltage (v)-Length Transduction (km)

Likewise, the voltage line required to cover the entire length range, from 0 to 80 km should be characterized- for 20 seconds of simulation- as follows

$$u(t)_{\to l(Km)} = \frac{80}{20} \cdot t = 20 \cdot t \qquad (25)$$

Underground Network Simulation

The simulation results of the 4 developed cases of underground distribution for the three analyzed networks are summarized in Table 4. To show the ability to adapt of the generated models, two graphs generated during the simulation process of the underground network conditions, Figures 5 and 6, are also included. Figure 5 shows the power loss progression associated with the Joule effect, which occurs in each of the networks (HVDCGR, HVDRMR, and HVAC) for a temperature of 25ºC, a total length of 80 km according to irradiance every step of the way, from 50 to 1000 (W/m^2).

The power loss progression is of quadratic nature since the losses are proportional to the square of the network's current, and it increases linearly with the increase of irradiance.

Figure 6 shows the progression of diminished capacity in the HVAC network, for a temperature of 25°C, a total length of 80 km and an irradiance of 1000 (W/m^2).

The graph shows a significant peak of power demanded at the beginning, which is necessary to charge the lines, diminishing its capacity due to the reactive energy required to charge/discharge both the capacitance of the lines and its inductance.

Aerial Network Simulation

The simulation results of the 4 developed cases of aerial distribution for the three analyzed networks are in Table 4. This case includes two graphs generated during the simulation process of the aerial network (see Figures 7 and 8).

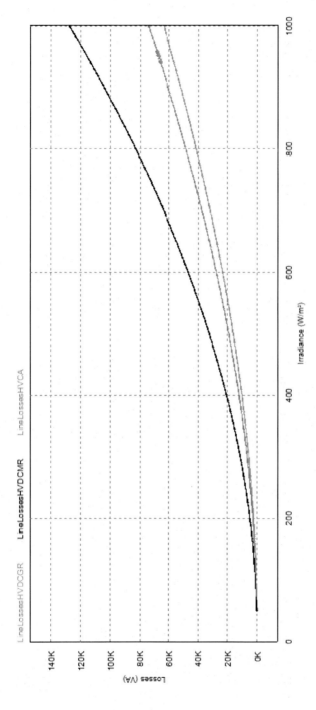

Figure 5. Losses due to Joule effect according to Irradiance, T° = 25°C, l = 80km, underground (self-developed).

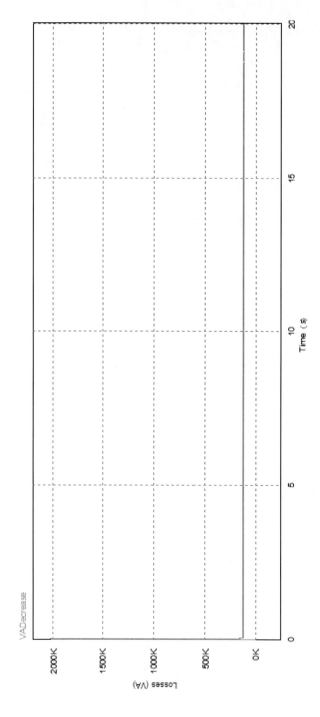

Figure 6. Decrease in line capacity HVAC under STC, I = (1000W/m^2), T$^{\circ}$ = 25°C and l = 80 km underground (self-developed).

Table 4. Transmit losses for each type of linK (self-developed)

Case U:UND; A:AIR T* (°C)/I (Wm2)	HVDC Groud Return		HVDC Metallic Return		HVAC 2-Phase			
	Joule Losses		Joule Losses		Joule Losses		Decrease Capability	
	J L (W)	J L (%)	J L (W)	J L (%)	J L (W)	J L (%)	DC (VA)	DC (%)
Underground Network: Insulated cables (Aluminum, S = 500mm2, XL = 0.095(Ω/km), C = 484(μF/km)								
U -10/1000	83,304.03	4.17%	166,608.06	8.33%	96,541.92	4.83%	161,498.98	8.07%
*U25/1000	64,001.00	3.20%	128,002.39	6.40%	74,599.95	3.73%	124,788.99	6.24%
**U48/800	34,240.00	1.71%	68,480.00	3.42%	40,350.05	2.02%	67,496.58	3.37%
U70/1000	42,409.77	2.12%	84,819.54	4.24%	48,809.60	2.44%	83,320.29	4.17%
Aerial Network: Cables without isolation (Aluminum, S = 500mm2, XL=0.35(Ω/km), C≈0								
A -10/1000	83,304.03	4.17%	166,608.06	8.33%	87,328.48	4.37%	440,180.00	22.01%
*A25/1000	64,001.00	3.20%	128,002.39	6.40%	67,592.28	3.38%	340,700.00	17.04%
**A48/800	34,240.00	1.71%	68,480.00	3.42%	36,647.89	1.83%	184,724.40	9.24%
A70/1000	42,409.77	2.12%	84,819.54	4.24%	45,210.80	2.28%	227,885.92	11.39%

Figures 5 to 8 belong to the shaded cases. *STC Standard test conditions, **NOCT Normal Operating Cell Temperature.

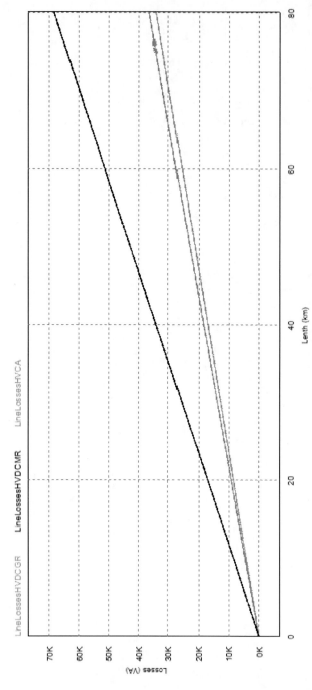

Figure 7. Joule effect losses as a function of length, aerial network and NOCT, T° = 48°C, I = 800 (W/m²) (self-developed).

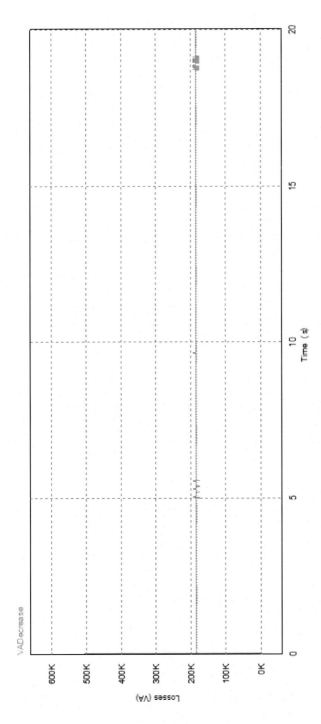

Figure 8. Decrease in line capacity HVAC, aerial network and NOCT, T° = 48°C, I = 800 (W/m²) (self-developed).

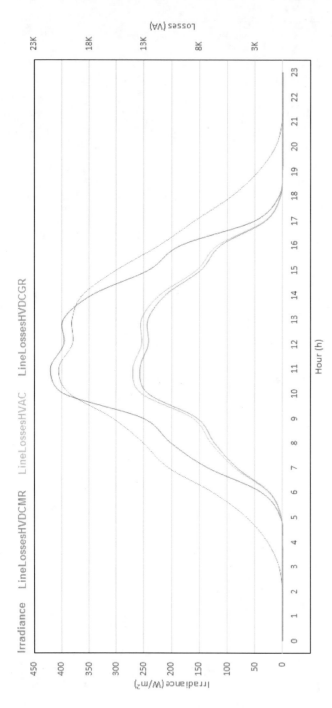

Figure 9. Losses due to Joule effect according to hour, underground, in the month of July in the Shetland Islands (self-developed).

Figure 7 shows the power loss progression, associated with the Joule effect, given in each of the networks (HVDCGR, HVDCMR, and HVAC) for a temperature of 48°C, an irradiance of 800 (W/m^2) according to length variation from 0 to 80 km. The graph shows the progression of such losses, which is of quadratic nature since the losses are proportional to the lines' ohmic effect, and this one increases proportionally to the length.

Figure 8 shows the progression of diminished capacity in an HVAC aerial network and for a temperature of 48°C in the cell, a total length of 80 km and an irradiance of 800 (W/m^2). Once again, the graph shows a significant peak of power demanded at the beginning, which is necessary to charge the lines, diminishing its capacity due to the reactive energy required to charge/discharge both the capacitance of the lines and its inductance.

SUMMARY RESULTS

Table 4 shows the summary of the results obtained in each of the four cases through a series of simulations, including the underground and aerial distribution networks, considering the total length for all three network types. These scenarios were selected because these are particularly significant in the photovoltaic power generation field of study.

From the results obtained, and given that an equal conductor cross section was used for the three types of networks in both types of distribution (500 mm^2), the following evaluations can be made;

- The network with fewest losses is the HVDCGR network since the link uses one single conductor. The losses in the HVDCMR system are of double digits in all cases, which is logical taking since the link has two conductors.

- The link with the most losses due to Joule effect is the HVDCMR, which is logical taking into account that such link carries all the current along two conductors and the HVAC the current is distributed through each phase in a relation less than $\sqrt{3}$.

- Joule losses in HVDCGR and HVAC are very similar, with greater differences in the case of underground distribution. The following should be noted:
  - In a real situation and due to the skin effect, the resistance of the phases in AC will be somewhat higher; therefore, one may think the difference between them is due to this effect. However, and given that the models for such resistive effects −of both networks− have been characterized in the same manner, with a nonlinear resistance which value increases with length, the skin effect is ignored.

  - After this simplification, and given that in each phase a $\sqrt{3}$ smaller current is circulating and the total losses are the sum of the three phases and, in the absence of capacitive and inductive phenomena, the losses in both cases must be equal:

$$JL_{HVDCGR} = R_{LINE} \cdot I^2_{PV\ to\ NETWORK} \approx JL_{HVAC} = 3 \cdot R_{LINE} \cdot \left(\frac{I_{PV\ to\ Network}}{\sqrt{3}} \right)^2$$

(26)

If these are not equal, it is because of the capacitive and inductive effects of the lines. This difference is more noticeable in underground networks, with a high capacitance. The constant loading/unloading on the lines increases the current on the line, experiencing a greater transmission loss. In the aerial line case-having neglected this effect and given a shorter line- the effect is smaller, and the losses are similar.

- In addition to the Joule effect losses, HVAC networks experience transmission losses due to a reduction in capacity, moreover HVAC aerial networks have an inductive reactance, 10 times greater, which is supporting evidence that the losses due to the Joule effect- decrease capacity lead to higher transmission losses in HVAC to those in the HVDC network.

- Both types of losses involve a financial cost: Joule losses represent a direct economic cost and capacity reductions indirect costs of under-utilization, which is necessary to capitalize on as operating expenses.

The best and most efficient option in the transport of production of the 2 MW PV plant is the use of an HVDC GR networks as it experiences the lowest losses of all and it only requires one conductor, followed by HVDC with a metallic return, which have a lower transit loss (collectively) and uses one less conductor. These findings are not only insightful, but are also a fundamental to objective of this study–the development of a simulation model for multi-megawatt photovoltaic plant and its different technologies HVAC and HVDC– to precisely detect transmit losses according to temperature and irradiance, with a certain network length.

Along with the simulations carried out, a Joule loss analysis was carried out for the three network types underground with real data of the indicated location, the Shetland Islands, with irradiation data obtained from [26]. The selected data are those of the average irradiance of the month of July, with the objective of having significant data since the Shetland Islands are slightly above 60 ° N of latitude. The results are shown in Figure 9. These findings culminate this research by drawing four very distinct, yet related conclusions. Findings have been compared with those available in top international journals [27-32] for comparative and validity purposes.

So far, little has been written in the scientific literature (in top international journals) about modeling multi-megawatt photovoltaic plants. To our knowledge, state-of-the-art of previous modeling of multi-megawatt photovoltaic plants has limited to using a wavelet variability model (WVM) for simulating solar photovoltaic (PV) power plant outputs (given a single irradiance point sensor time series using spatio-temporal correlations). See, for example, Lave et al. [33]; or to propose new control strategies that enable the storage requirements to smooth out short-term power fluctuations. See, for example, de la Parra et al. [34]. As a consequence, authors can reasonably prove that this chapter incorporates some features that falls outside the pool of existing knowledge of its technical realm, which makes this chapter particularly interesting.

CONCLUSION

The first conclusion relates to the generated models used in this research. Such models have adequately measured the transmission losses for any combination of the variables according to temperature, irradiance, and with a certain network length by evaluating and comparing the viable transport distances for HVAC, HVDC while allowing the length adjustment of said links. Moreover, these models have quantified performance variation in MMW PV plants at different temperatures; critical to accurately determining the increase in production as well as the losses incurred from changes in irradiation by applying irradiance variation's profile and adjusting the time during simulation, obtaining the total energy loss (Wh). Another key aspect of these models is the development of a tool that allows capitalizing the operating costs of the PV plant networks based on a multitude of combinations of said variables, simplifying the calculation of the various links' critical distance, thus making this process more efficient. Last, the modular nature of these models offers scalability for other PV plant configurations while making its parametrization simple as evident in the case study.

Next is technical feasibility of MMV PV plant integration in HDVC networks. The distribution of energy produced in MMW PV plants via HVDC networks is technically viable. These networks would work in the 15 to 30 kV range and power in the order of MW, which is considered a small-scale HVDC network.

Given the evidence presented in this chapter, the best alternative for electrical distribution are HVDC networks a ground return as the first choice and metallic return as the second option, due to the decrease of capacity of the HVAC networks. Regarding the efficiency of the PV field converter unit, the most efficient one is the combination of the existing medium voltage multi-megawatt PV with the addition of a final rectifier stage, as a small-scale HVDC converter station. This last state – given the power and voltage requirements– can consist of a multi-level structure IGBT type switches given the ranges of medium voltage output voltages required by the links. The last point involves transmission losses; therefore, and given both the thermal losses of the Joule effect and diminished capacity present in HVAC networks– resulting in line availability– it is concluded that, as a whole, the HVAC lines experience greater losses than the HVDCGR. Thus, the smaller scale environment PV plants operate under is more convenient to transport via the HVDC network.

The answer to the economic feasibility of the use of MMW PV plants in HVDC networks is that for HVDC transport to be economically viable, the cost between the two transport options must level. It will happen when the savings resulting from the reduction of conductors and transit losses equals that of the multilevel rectifier unit plus the cost of the end-of-line converter station, which is becoming increasingly feasible because of substantial cost reductions in power electronics and the small-scale power being handled.

The last differentiated yet related conclusion relates to MMW PV applications via HDVC networks. The output produced by -MMW PV integrated into small scale HVDC networks- offers a myriad of possibilities including market penetration. One of these involves the use of floating PV plants in reservoirs or islands close to the shore, using one single conductor, decreasing losses as already demonstrated. Another possibility has the potential to become a very viable alternative to increase the electrification rate of developing countries, given the reduction of costs that can be achieved, the large transport distances that can be reached for distributed systems and the great solar capital available in the majority of these.

APPENDICES

Appendix A. HVDC Network Modeling

There are different interconnection topologies between the converter stations in HVDC networks. Given the complexity of the multi-terminal control, the most common use, as already indicated, it is the point-to-point interconnection. The number of cables interconnecting stations varies. The most common are;

- Monopolar links with *Ground Return, HVDCGR*. It uses a single conductor to transmit electrical energy. The return uses connected electrodes, which are in turn connected to the conversion stations. These electrodes perform the functions of anode and cathode. This type of connection connects systems separated by long distances and where the non-installation of the return cable can lead to considerable savings [2].

- Homopolar links with a metallic return, *HVDCMR*. A few monopolar systems include a metallic return when grounded electrodes are not an option, typically because of environmental issues, or when losses are significant [2].
- Bipolar links are currently used the most for its greater power transmission capacity.

Ground return monopole systems are simpler and cheaper than other systems and are best suited for moderate energy transfer [4]. Therefore, in the case of MWPV plants, the most suitable link is monopolar, given the small-scale level of the power to be transmitted.

Appendix B. HVAC Network Modeling

When modeling AC lines of medium to long length or in all distances when it comes to AC underground lines, it is important to consider three factors. These include the effects exerted by both the resistance and the inductance, and most importantly, the effect of capacity; responsible for leakage currents, capable of decreasing transport capacity until it is rendered unfeasible. The T-equivalent circuit method or the "Pi" method can be used to consider these effects [22].

The "Pi" method is suitable as it provides greater versatility in simulations, see Figure 34, with the half capacitances connected in parallel through two manual switches S1 and S2. These switches are connected according to the length of the line, that is, depending on whether or not the capacitive effects can be ignored.

For the characterization of these parameters, it is necessary to consider:

- The ohmic value determines the resistance for each phase, and it includes the variable length, just as in the case of the HVDCGR or HVDCMR. Because of the skin effect present in AC, there is more current flowing around the periphery of the conductor, so the total resistive value of the line is higher, but this effect has no relevance in low frequencies such as 500.95 work of the link.

- Despite conductance's (G) ability to measure the leakage current of both the insulators and those due to the corona discharge, it is not used in many case studies because conductance's incidence is low under normal operations conditions and is complicated to calculate. As a result, it is often disregarded and considered to be infinite.

- The total inductive reactance value of each phase is estimated following II.F.

- The manufacturer's data is needed to determine the value of the network capacity required for the model. The data includes isolated conductors, and it is calculated according to the design of aerial networks with bare conductors as previously mentioned.

REFERENCES

[1] Francescutti P. *La guerra de las Corrientes*. [The War of the Currents.] Available from: http://www.ree.es/sala_prensa/ext_img/entrelineas-0007_5.pdf [Accessed December 28th 2017].

[2] Bahrman M and Johnson B. The ABCs of HVDC transmission technologies. *IEEE Power and Energy Magazine*. 2007;52(2):32–44.

[3] Moglestue A. *60 years of HVDC. ABB Review*. 2014; (2):33–41.

[4] Red Eléctrica de España. *Interconexión subterránea España-Francia* [*Spain-France underground interconnection*]. Available from   http://www.ree.es/es/actividades/proyectos-singulares/nueva-interconexion-electrica-con-francia   [Accessed December 28th 2017].

[5] Ohlsson D, Korbel, J, Steiger U. Opening move. *ABB Review.* 2013; (3):27–33.

[6] Llamas D. *Proyecto para llevar energía termosolar desde Sáhara Europa* [*Project to bring solar thermal energy from Sahara Europe*] Available from: http://helionoticias.es/proyecto-para-llevar-energia-termosolar-desde-el-sahara-hasta-europa/ [Accessed December 28th 2017].

[7] Ministerio de Economia y competitividad. Congreso Iberoamericano sobre Microrredes con Generación Distribuida de Renovables [Ibero-American Congress on Microgrids with Distributed Generation of Renewables], October 11-13, 2012, *Centro de Investigaciones Energeticas Medioambientales y technologicas*, Soria, España; 2012.

[8] Ruiz P. *Se inauguran en Japón dos grandes centrales solares flotantes* [*Japan inaugurates two large floating solar power plants*]. Available from: https://www.energias-renovables.com/fotovoltaica/se-inauguran-en-japon-dos-grandes-centrales-20150519 [Accessed 28th July 2017].

[9] *XVI Congreso Internacional de Ingenieria de Proyectos.* Cubrición de embalses mediante un sistema de cubierta flotante fotovoltaico: Análisis técnico y económico [*International Congress of Project Engineering.* Covering reservoirs using a photovoltaic floating roof system: Technical and economic analysis], July 11-13, 2014, Universidad Politecnica de Valencia, Valencia, España; 2014.

[10] Serrano A. *El nuevo reto de la Energía Solar* [*The new challenge of solar energy*] Available from: http://www.imf-formacion.

com/blog/energias-renovables/articulos/energias-renovables-articluos/energia-solar/ [Accessed: December 28th 2017].

[11] Frau J, Gutiérrez J, Transporte de energía eléctrica en corriente continua: HVDC. Estado actual y perspectivas [Transmission of electrical energy in direct current: HVDC. Current status and prospects]. *Electrónica De Potencia, automática e instrumentación.* 2005;(361): 2–14.

[12] Mohan N, Undeland T, Robbins W. *Electrónica de potencia: convertidores, aplicaciones y diseño* [Power electronics: converters, applications and design] 3th ed. Mexico. McGraw-Hill; 2009.

[13] Salas Merino V, Olías Ruiz E, Grupo de Sistemas Electrónicos de Potencia, Departamento de Tecnología Electrónica, Universidad Carlos III de Madrid. *Análisis de las grandes estaciones de potencia fotovoltaicas (multi megavatios) de media tension,* 2013 [Analysis of large photovoltaic (multi-megawatt) medium voltage power stations]. Available from: http://www.jemaenergy.com/ images/about/press/Instalaciones-en-grandes-plantas-foto voltaicas-de-conexion-a-red.pdf [Accessed: December 29th 2017].

[14] Saavedra Montes A, Ramos C, and Trejos Grisales L. Adaptive maximum power point tracking algorithm for multi-variable applications in photovoltaic arrays. *Journal EIA.* 2013; 10 (20): 193–206. Available from: http://www.scielo.org.co/pdf/ eia/n20/n20a17.pdf [Accessed: December 29th 2017].

[15] Gonzalez-Longatt F. *Modulación Por Ancho de Pulso.* [*Pulse Width Modulation*] 2004. Available from: http://fglongatt.org/ OLD/Reportes/PRT2004-02.pdf [Accessed: December 29th 2017].

[16] Bueno E, Garcia R, Marrón M, and Urena J. et al. Modulation Techniques Comparison for Three Levels VSI Converters. In: *28th Annual Conference IEEE Industrial Electronics Society.*

Sevilla (Spain). IEEE. 20 March 2003. pp. 908–913. Available from: DOI 10.1109/IECON.2002.1185393.

[17] Rashwan M. State of the VSC technology. [Presentation] *IEEE 2011 Electrical Power and Energy Conference.* October 2011.

[18] Powersim Software AS. *PSIM User's Guide.* Version 11.0. May 2017.

[19] Powersim Software AS. *Solar Module Physical Model Tutorial.* Version 11.0. October 2016.

[20] *Datasheet NU-RC290 | 290 W, SHARP.* Available from: https://eng.sfe-solar.com/wp-content/uploads/2016/03/SHARP_NURC290W_Mono_EN.pdf [Accessed: December 29th 2017].

[21] Fernandez H. *Convertidores de potencia Aplicaciones y Análisis con el PSIM [Power Converters Applications and Analysis with the PSIM]* Madrid: Kindle; 2014.

[22] Mujal-Rosas R. *Cálculo de líneas y redes eléctricas. [Calculation of electric lines and networks.]* Barcelona: Ediciones de la Universidad Politécnica de Catalunya; 2002.

[23] Prysmian Spain, S.A. *Cables y Accesorios para Media Tensión. [Cables and Accessories for Medium Voltage.]* July 2014.

[24] *Database of weather in the UK.* http://www.weather.org.uk/climate/scotclim.html [Accessed: December 29th 2017].

[25] Fayos-Alvarez A. *Líneas eléctricas y transporte de energía eléctrica.* Valencia: Universidad Politécnica de Valencia; 2013.

[26] *World Radiation Data Centre.* http://wrdc.mgo.rssi.ru/wrdc_en_new.htm [Accessed: December 29th 2017].

[27] Kalair A, Abas N, Kalair AR, Saleem Z, Khan N. Review of harmonic analysis, modeling and mitigation techniques. *Renewable and Sustainable Energy Reviews* 2017;78:1152–1187.

[28] Aliyu AK, Modu B, Tan CW. A review of renewable energy development in Africa: A focus in South Africa, Egypt and Nigeria. *Renewable and Sustainable Energy Reviews* 2018; 81(2):2502–2518.

[29] IqtiyaniIlham N, Hasanuzzaman M, Hosenuzzaman M. European smart grid prospects, policies, and challenges. *Renewable and Sustainable Energy Reviews* 2017;67:776–790.

[30] Dadhania A, Venkatesh B, Nassif AB, Sood VK. Modeling of doubly fed induction generators for distribution system power flow analysis. *International Journal of Electrical Power & Energy Systems* 2013;53:576–583.

[31] Brennand TP. Renewable energy in the United Kingdom: policies and prospects. *Energy for Sustainable Development* 2004;8(1):82–92.

[32] Lior N. Sustainable energy development: The present (2009) situation and possible paths to the future. *Energy* 2010; 35(10):3976–3994.

[33] Lave M, Kleissl J, Stein JS. A wavelet-based variability model (WVM) for solar PV power plants. *IEEE Transactions on Sustainable Energy* 2013;4(2):501–509.

[34] de la Parra I, Marcos J, García M, Marroyo L. Control strategies to use the minimum energy storage requirement for PV power ramp-rate control. *Solar Energy* 2015;111:332–343.

In: Solar Power Technology   ISBN: 978-1-53614-204-4
Editors: A. Colmenar-Santos et al. © 2019 Nova Science Publishers, Inc.

Chapter 3

PV GRID CONNECTED
INVERTERS SIMULATION

Luis Dávila-Gómez[1,] and Enrique-Luis Molina-Ibález[2]*
Departamento de Ingeniería Eléctrica, Electrónica, Control,
Telemática y Química Aplicada a la Ingeniería, Universidad
Nacional de Educación a Distancia (UNED), Madrid, Spain

ABSTRACT

This chapter proposes a new model for characterizing the energetic behavior of grid connected PV inverters. The model has been obtained from a detailed study of main loss processes in small size PV inverters in the market. The main advantage of the used method is to obtain a model that comprises two antagonistic features, since both are simple, easy to compute and apply, and accurate. One of the main features of this model is how it handles the maximum power point tracking (MPPT) and the efficiency: in both parts the model uses the same approach and it is achieved by two resistive elements which simulate the losses inherent to each parameter. This makes this model easy to implement, compact and refined. The model presented here also includes other parameters, such as

---

* Corresponding Author Email: l.davil.lopez@gmail.com.

start threshold, standby consumption and islanding behavior. In order to validate the model, the values of all the parameters listed above have been obtained and adjusted using field measurements for several commercial inverters, and the behavior of the model applied to a particular inverter has been compared with real data under different working conditions, taken from a facility located in Madrid. The results show a good fit between the model values and the real data. As an example, the model has been implemented in PSPICE electronic simulator, and this approach has been used to teach grid-connected PV systems. The use of this model for the maintenance of working PV facilities is also shown.

Keywords: grid-connected photovoltaic systems, empirical models, efficiency, parameter extraction, modeling

INTRODUCTION

Simulation of photovoltaic systems is becoming more and more important. There are a wide variety of software packages that allow you to make easier the processes of design and analysis of grid connected photovoltaic systems.

Referring to those focused on the design, we can find tools for the economic calculation of the systems, tools for dimensioning of the components, and those that integrate both economics and sizing issues.

The software programs for economic analysis allow you to calculate the total cost of photovoltaics systems, because they contain a database of commercial elements and their market prices. If the software includes the calculation of the energy generated, it will allow us to determine the payback time of the installation and thus calculate investment feasibility. These programs perform relatively simple calculations, and require little interaction with the user. In this group we can find RETScreen [1] and PVSYST [2].

For the sizing of photovoltaic systems, some companies have developed their own programs, mostly in spreadsheets, which allow them to undertake their projects of PV systems, but there are other programs that perform this function [3].

Referring to those focused on the analysis, we found that some of them simulate the generation of the system, others estimate shading losses, others help to optimize the system, etc.

PV simulation programs take data from a system already sized and provide a detailed time analysis of its operation. This analysis allows us to verify the sizing done, to check the impact on production of a malfunction or a voluntary stop of the system, to test the performance under different conditions, etc. Among others, the most used are: HOMER [4], which allows the simulation of hybrid systems, PV-DesignPro [5], and PVSYST.

On the other hand, generic simulators are those designed for the study of different types of problems. Most of them also allow simulations of photovoltaic systems, both stand-alone and grid-connected. There are a wide variety of tools of this type, and there have been published papers where you can find examples of applications in photovoltaic systems; for example with the programming languages R [6] or JAVA [7], or in commercial programs for electrical simulation [8-10]. From those used for photovoltaic simulation we can highlight two:

- SIMULINK is suitable when you want to simulate the behavior of systems with photovoltaic and mechanical parts such as an electric vehicle, as in [11]. But it may also be applied to the simulation of control strategies in photovoltaic systems [12], to the performance evaluation of the MPPT of the inverter [13, 14], to the improvement in the design of PV inverters [16] and to simulate the behavior of PV cells and modules [17, 18]. SIMULINK is also useful for full modeling of stand-alone, grid-connected and hybrid photovoltaic systems [19], and allows identifying the system using parameters extraction

methods from real measurements and diagnosing potential
faults [20], which helps in the maintenance of the systems. We
also found examples of application in the area of the
photovoltaic systems teaching in [21].

- SPICE simulation environment is the mostly used in
  electronics. Also its use for the simulation of photovoltaic
  systems has a long history. There are a vast literature on power
  converters simulations [22, 23], the study of their faults [24]
  and the MPPT [25] to validate its use in photovoltaic systems.
  In regard to the photovoltaic elements, we can find models of
  the solar cell [26] and the solar panel [27], these models can
  help to study the effect of shading and propose solutions [28], to
  determine the need for the use of bypass and blocking diodes
  [29,30]. In academia and for teaching purposes the simulator
  was used to study the performance of solar cells and panels [26]
  and there are tools to simulate photovoltaic systems in
  Undergraduate and Master Degree courses [31].

There are two different kinds of models to simulate a photovoltaic
inverter:

1. Topological models, which includes the whole electric scheme
   of the inverter, or a simplified version of it, as in [8-10, 12, 15,
   16, 23, 24, 31]. These models allow detailed short time
   simulations and they are very useful for the electronics design.
   However, usually the models are too complex and only can
   simulate the inverters for which they were created, and have a
   little use in PV simulation. They are more concerned with the
   long-term behavior of the system. These models are not related
   to the model proposed in this chapter, thus it will not be studied.
2. Behavioral models, which are focused on the input-output
   relationships and on the working principles of the inverter.

These models are more generics and can simulate any inverter with the same operating principle. There are application examples in [22] for islanding behavior, for the maximum power point tracking (MPPT) in [13, 14, 25, 31], and for the inverter efficiency in [25, 31, 32].

The proposed model belongs to the group "behavioral models"; for this reason some of them are detailed next.

The MPPT model in [13] simulates the control strategy, which is different for every case of study. To simulate the efficiency, it uses only six values of the efficiency curve. A model is proposed in [14] with two main blocks: a DC/DC converter and a control circuit. In [25], a digital controller is developed for a MPPT based on "Perturbation & Observation" method. Finally, [31] shows a model based on an ideal DC/DC converter with constant MPPT efficiency.

The efficiency model proposed in [25] is a lookup table with values depending on the input power. For intermediate values, a linear interpolation is used. The model presented in [31] considers a constant efficiency for the inverter, so the output power is oversized for low values of the input power. The one proposed in [32] is a losses model, similar to the proposed in this chapter. Both will be compared in section "3.2 Model Validation".

This chapter focuses on the inverter of photovoltaic systems and their characterization. The rest of the chapter is structured as follows. The second section describes a model for photovoltaic inverter that can be simulated with any of the simulators described above. The third section presents the methodology for the calculation and validation of the model. The fourth section shows a selection of examples based on the last application mentioned in the previous paragraph.

Finally, the drawn conclusions from this contribution are provided.

INVERTER MODEL

The inverter model developed can simulate several characteristics related with its behavior. These can be divided into two groups: Energetic performance and islanding protections.

Energetic Performance Characteristics

In the energetic performance group, the mains elements are:

1. Efficiency, the ratio between AC output power and DC input power.

$$\eta = \frac{P_{AC}}{P_{DC}} \tag{1}$$

2. Start threshold; it is the minimum DC power needed to start the power conversion.
3. Standby consumption is the power wasted by the inverter when no power conversion is making.
4. Maximum power point tracking. The inverter circuits try to obtain all power of PV arrays, and this feature measures how near it is. It can be modelled as a new efficiency, and can be defined as the ratio between power obtained from the PV array by the inverter and theoretical maximum power available from the PV array.

$$\eta_{MPPT} = \frac{\int P_{REAL}\, dt}{\int P_{MAX}\, dt} \tag{2}$$

Energetic Performance Models

A model has been developed that simulates the inverter efficiency based on inverter losses. Unlike behavioral models [25, 31, 32], this model is based on the one developed by N. Chivelet [33, 34] for stand-alone inverters. The model has been adapted here for grid connected inverters.

We can expect two kinds of losses in an inverter:

1. Losses due to the consumption of inverter circuits, those always are present when the power conversion is on. These losses are independent on input power.
2. Losses that depend on the amount of power conversion (commutation losses, wiring losses). These losses are strongly dependent on input power.

The action of both types of losses shapes the efficiency curve.

The model is made with an ideal inverter (the one with efficiency equal to one for all possible power output values), which has a serial input resistor and a parallel output resistor.

According to previous model description, the electrical scheme of this approach would be as shown in Figure 1.

The efficiency can be calculated as follows:

$$\eta = \frac{P_{AC}}{P_{DC}} = \frac{V_{AC}I_{AC}}{V_{DC}I_{DC}} \tag{3}$$

To write the efficiency as function of input voltage and power output, you must get the value of I_{DC}. In order to get I_{DC}, you must take into consideration that in an ideal inverter power input equals to power output:

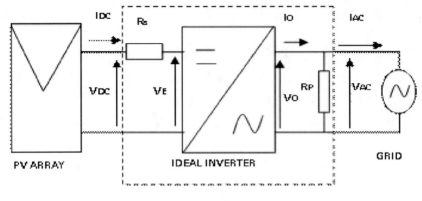

Figure 1. Equivalent electric circuit of a grid connected inverter.

$$V_E I_{DC} = V_0 I_0 = V_{AC} I_0 \tag{4}$$

This equation can be written as follows:

$$V_{DC} I_{DC} - I_{DC}^2 R_S = \frac{V_{AC}^2}{R_P} + V_{AC} I_{AC} \tag{5}$$

$$R_S I_{DC}^2 - V_{DC} I_{DC} + \left(\frac{V_{AC}^2}{R_P} + P_{AC} \right) = 0 \tag{6}$$

Solving for DC current:

$$I_{DC} = \frac{V_{DC} \pm \sqrt{V_{DC}^2 - 4 R_S \left(\frac{V_{AC}^2}{R_P} + P_{AC} \right)}}{2 R_S} \tag{7}$$

From the two possible solutions, we took the one with negative sign before the square root, and calculating we obtain:

$$I_{DC} = \frac{V_{DC} - V_{DC}\sqrt{1 - 4\frac{R_S}{V_{DC}^2}\left(\frac{V_{AC}^2}{R_P} + P_{AC}\right)}}{2R_S} \tag{8}$$

That introduced in Equation (3) gives us:

$$\eta = \frac{P_{AC}}{V_{DC}I_{DC}} = \frac{P_{AC}}{V_{DC}\left[\dfrac{V_{DC} - V_{DC}\sqrt{1 - 4\dfrac{R_S}{V_{DC}^2}\left(\dfrac{V_{AC}^2}{R_P} + P_{AC}\right)}}{2R_S}\right]} \tag{9}$$

Thus, the final equation for the efficiency is as follows:

$$\eta = \frac{2R_S P_{AC}}{V_{DC}^2\left(1 - \sqrt{1 - 4\dfrac{R_S}{V_{DC}^2}\left(\dfrac{V_{AC}^2}{R_P} + P_{AC}\right)}\right)} \tag{10}$$

Equation (10) gives us the efficiency as a function of DC input voltage, RMS AC output voltage and output power, and includes the two resistive parameters: R_S and R_P. These resistive parameters represent the inverter losses.

R_S stands for coupling losses between DC and AC side of the inverter. Figure 2 shows the variation of the efficiency when the value of this parameter is modified. In the figure, the more value of R_S, the less value of efficiency in the flat zone of the graphic, and as you can see, the shape of the graphic is not affected by the variation of this parameter. R_S is a serial resistor placed on the input so that power losses due to this component increases when input power increases. Thus, the higher the output power, the higher the contribution of R_S to efficiency. Thereby, R_S is related to the losses that depend on the input power, such as commutation losses.

The R_p parameter is responsible of the low power zone shape, as we can see in Figure 3. As you can see, the higher the value of R_p, the higher the slope of the curve, and if R_p is high enough the efficiency becomes almost ideal ($\eta = 1$ for all possible power output values). R_p is a parallel resistor in the output stage of the inverter; power losses on this component are proportional to the square of RMS AC output voltage. Output voltage is equal to grid voltage, and then loss on R_p will be a constant quantity over all the output power range. Then, R_P represents operation losses of inverter. This means that the influence of R_p will be very important in low power zone of efficiency, decreasing as power output increases.

The inverter efficiency does not have valid values below the start threshold. The curves obtained with the efficiency model have values not equal to zero when output power is near to zero because the model does not include the start threshold characteristic of the inverter. However, the start threshold can be easily included if you modify Equation (10) in this way:

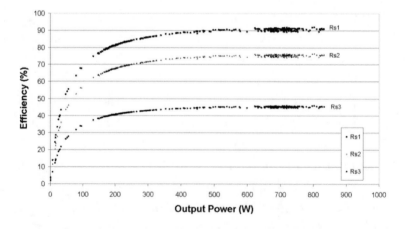

Figure 2. Efficiency variation as a function of RS parameter (RS1 < RS2 < RS3).

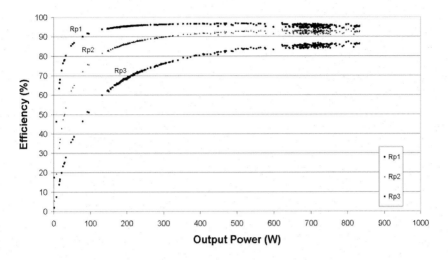

Figure 3. Efficiency variation as a function of RP parameter (RP1 > RP2 > RP3).

$$\eta = 0 \qquad if \ P_{DC} < P_A \tag{11}$$

$$\eta = \frac{2R_S P_{AC}}{V_{DC}^2 \left(1 - \sqrt{1 - 4\frac{R_S}{V_{DC}^2}\left(\frac{V_{AC}^2}{R_P} + P_{AC}\right)}\right)} \quad if \ P_{DC} > P_A \tag{12}$$

Were P_A is the input power necessary to the inverter to start the conversion.

The efficiency model proposed in this work is different from those pointed in the Introduction. In [13, 25] the models are limited to a few specific values, while the proposed can cover the whole curve. In [31] the efficiency is a constant; in contrast, in the proposed model the efficiency varies with inverter parameters. Finally, the model from Jantsch et al. [32] is the most similar to the proposed, but it needs three parameters instead of the two resistive parameters presented here. This model will be compared with the proposed in section "3. Model Calculation."

When the inverter stops energizing the utility (for example, when the DC power remains below the start threshold), there are an amount of power consumption from the grid, that is called the standby consumption. To model this characteristic, you can use a resistor R_{SC} to simulate this power like a loss. Then, the value of R_{SC} will be as follows:

$$R_{SC} = \frac{V_{AC}^2}{P_{SC}} \qquad (13)$$

Where P_{SC} is the power needed in the standby mode of the inverter.

This resistor will replace the parallel output resistor R_P when the inverter enters in standby mode, while efficiency is equal to zero. This part of the model is shown in Figure 4.

The last characteristic for the model is the maximum power tracking. The parameter we will use to determine the performance of a maximum power point tracker is the efficiency [14], as in equation (2). But if we define instantaneous power efficiency as in [35]:

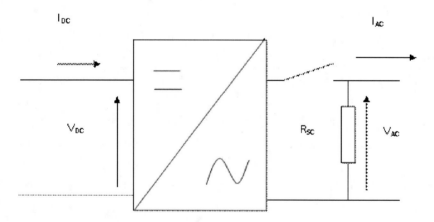

Figure 4. Standby consumption part of the model.

$$\eta_{MPPT} = \frac{P_{DC}}{P_{MAX}} \tag{14}$$

This efficiency depends on the input power, and you can find two different losses in this parameter:

- On the one hand, low power losses, which are relatively high due to the difficulty in determining the optimum operating point when the power curve is very flat.
- Moreover, for high power values the tracker has very small losses, since the perfect track does not exist.

The combination of both types of losses makes the instantaneous power efficiency curve. This one is very similar to the efficiency curve of the inverter, and is simulated using a similar model.

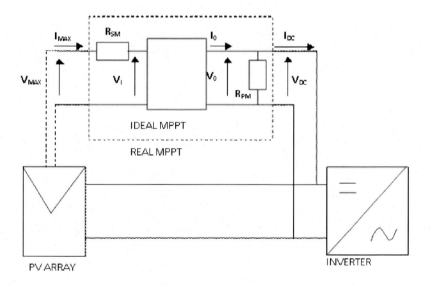

Figure 5. Equivalent electric circuit of the MPPT of a grid connected inverter.

We will simulate MPPT losses using two resistors, one in series with the input and another in parallel with the output of an ideal MPPT, without any loss. This is a new model with the physical sense of a loss model, as opposed to purely mathematical models [36, 37] or those based on the topology of the electronic circuitry [13] or its behavior [38]. According to the description of the model, the scheme would be the one shown in Figure 5.

The MPPT efficiency can be calculated in the same way as inverter efficiency. Then, we can write Eq. (14) as follows:

$$\eta_{MPPT} = \frac{P_{DC}}{P_{MAX}} = \frac{V_{DC}I_{DC}}{V_{MAX}I_{MAX}} \tag{15}$$

For the ideal MPPT:

$$V_I I_{MAX} = V_0 I_0 = V_{DC} I_0 \tag{16}$$

To obtain I$_{MAX}$:

$$V_{MAX} I_{MAX} - I_{MAX}^2 R_{SM} = \frac{V_{DC}^2}{R_{PM}} + V_{DC} I_{DC} \tag{17}$$

$$R_{SM} I_{MAX}^2 - V_{MAX} I_{MAX} + \left(\frac{V_{DC}^2}{R_{PM}} + P_{DC} \right) = 0 \tag{18}$$

$$I_{MAX} = \frac{V_{MAX} \pm \sqrt{V_{MAX}^2 - 4R_{SM}\left(\frac{V_{DC}^2}{R_{PM}} + P_{DC} \right)}}{2R_{SM}} \tag{19}$$

From the two possible solutions, we took the one with negative sign before the square root, and calculating we obtain:

$$I_{MAX} = \frac{V_{MAX} - V_{MAX}\sqrt{1 - 4\frac{R_{SM}}{V_{MAX}^2}\left(\frac{V_{DC}^2}{R_{PM}} + P_{DC}\right)}}{2R_{SM}} \quad (20)$$

Thus, from Equation (15), the final formula for the MPPT efficiency can be written as follows:

$$\eta_{MPPT} = \frac{2R_{SM}P_{DC}}{V_{MAX}^2\left(1 - \sqrt{1 - 4\frac{R_{SM}}{V_{MAX}^2}\left(\frac{V_{DC}^2}{R_{PM}} + P_{DC}\right)}\right)} \quad (21)$$

Equation (15) includes two resistive parameters: R_{SM} and R_{PM}. R_{SM} represents the high power losses. R_{SM} is a resistor connected in series with the input so that the power loss on it increases with the value of the input power. Its contribution to the shape of the MPPT efficiency is the same as shown in Figure 2 for R_S.

R_{PM} is the parameter for low power losses. If R_{PM} value increases, the slope of the curve increases too, and for high values of the resistive parameter, the MPPT efficiency is near the ideal value ($\eta_{MPPT} = 1$ for any output power). The resistor is in parallel with the output; then the losses depend on the square of the output voltage. Since the output voltage varies mainly with the variation of temperature, the loss due to this resistor varies very slowly. Its influence on MPPT efficiency is very important in the low power zone, and the shape of the curve is modified in the same way as R_P in Figure 3.

The MPPT efficiency model proposed in this work is different from those pointed in the Introduction. Most models are based on simulating one control strategy [13, 14, 25] and then are only suitable for inverters using these control strategies. In [31], MPPT is a constant; in contrast to this new model, which has a MPPT efficiency that varies with inverter parameters.

Quality Characteristics

This set of characteristics evaluates if the signal is within the operating range. For the proposed model, we use the following:

1. Over voltage: it is the value of the utility voltage above a certain threshold.
2. Under voltage: by analogy, the under voltage is a voltage value on the utility below a set limit.

If there is one of these situations, the inverter should stop energizing the grid.

Quality Models

The model used to simulate these characteristics is very simple. We only make a comparison between RMS present value of grid voltage and the limits set according to standard IEC 61727. When the comparison shows an out of limit situation, we force the inverter to the standby mode.

Thus, the model will have two parameters: the lower limit of voltage, V_{min} and the upper one, V_{max}.

Electrical Security Characteristics

In our case, the most relevant feature is the islanding behavior. Islanding is a state in which a portion of the electric utility grid, containing load and generation, continues to operate isolated from the rest of the grid. This represents a potential safety hazard of a line worker coming in contact with an energized line that is presumed to be de-energized [39].

Any grid-connected inverter has to detect island conditions and should cease to energize the public electric power grid under specific conditions. This conditions and the test procedure are the purpose of the IEC 62116 standard. The inverters that meet the requirements of IEC 62116 and IEC 61727 can be considered non-islanding.

Electrical Security Model

In our model, to simulate the islanding behavior, we force the inverter to enter the standby mode when a loss of power occurs. This situation is detected with the quality model. The inverter shall not reconnect before the frequency and voltage have been maintained within the specified limits for at least the time addressed in IEC 61727.

Table 1. Parameters for the model

Characteristic	Parameter(s)
Energetic performance	
Efficiency	R_S, R_P
Start threshold	P_A
Standby consumption	R_{SC}
MPPT	R_{SM}, R_{PM}
Quality	
Over and Under Voltage	V_{min}, V_{max}
Electrical security	
Islanding behavior	T_R

So the model needs only one parameter: the reconnection time T_R.

The electrical security model and the quality model are interconnected, because both have to force the inverter to enter the standby mode. Therefore, both models generate a condition of "standby mode" that must be combined to generate the correct alarm.

Table 1 shows all the parameters required for the different parts of this model.

MODEL CALCULATION

To implement the inverter model developed, it is necessary to obtain the numerical values of the parameters for a particular device. Some of them can be found in its data sheet, or can be calculated from this information. Using the parameters calculated for several inverters we can validate our model.

Parameters Extraction

In Table 1 we have all the parameters for the model. The nominal values of many of them can be found in the data sheet of the inverter. We can find values for P_A, V_{min}, V_{max} and T_R. On the other hand, R_{SC} can be calculated from P_{SC}, included in the data sheet. But dissipative parameters are not included in data sheets, and must be calculated.

There are four dissipative parameters: R_S, R_P for the overall efficiency and R_{SM}, R_{PM} for the MPPT efficiency. Methods for calculating R_S and R_P are the same as for R_{SM} and R_{PM} because we use a similar approach.

To calculate R_S, we take Equation (5) and solve for this parameter:

$$R_S = \frac{V_{DC}I_{DC} - \dfrac{V_{AC}^2}{R_P} - V_{AC}I_{AC}}{I_{DC}^2} \quad (22)$$

When input and output powers are high, the R_P losses term can be ignored. Then:

$$R_S \approx \frac{P_{DC} - P_{AC}}{I_{DC}^2} \quad (23)$$

In order to evaluate R_S we must use a set of values for P_{DC}, P_{AC} and I_{DC} as higher as possible, between ones obtained in measures.

To evaluate R_P, we take again Equation (5) and find now R_P:

$$\frac{1}{R_P} = \frac{V_{DC}I_{DC} - I_{DC}^2 R_S - V_{AC}I_{AC}}{V_{AC}^2} \quad (24)$$

When input and output powers are low, the R_S losses term can be ignored this time. Then, we obtain:

$$\frac{1}{R_P} \approx \frac{P_{DC} - P_{AC}}{V_{AC}^2} \quad (25)$$

Several test showed that the most suitable set of values P_{DC}, P_{AC} and V_{DC} for calculating R_P are those nearest to the zone of slope change in efficiency curve [40].

Likewise, to calculate R_{SM}, we take now equation (17) and solving:

$$R_{SM} = \frac{V_{MAX}I_{MAX} - \dfrac{V_{DC}^2}{R_{PM}} - V_{DC}I_{DC}}{I_{MAX}^2} \tag{26}$$

And for high values of input and output power, the solution can be approximated to:

$$R_{SM} \approx \frac{P_{MAX} - P_{DC}}{I_{MAX}^2} \tag{27}$$

To evaluate R_{PM}, we take again equation (17) and we arrange it as follows:

$$\frac{1}{R_{PM}} = \frac{V_{MAX}I_{MAX} - I_{MAX}^2 R_{SM} - V_{DC}I_{DC}}{V_{DC}^2} \tag{28}$$

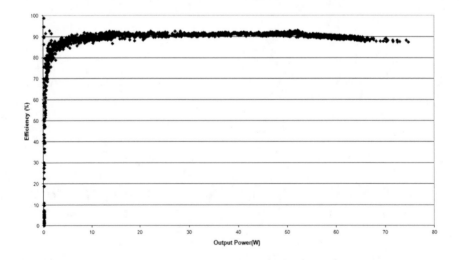

Figure 6. Measured curve of efficiency for a commercial inverter.

In the low power zone, the R_{SM} losses term can be ignored, and we obtain:

$$\frac{1}{R_{PM}} \approx \frac{P_{MAX} - P_{DC}}{V_{DC}^2} \tag{29}$$

As in the efficiency sub model, the most suitable set of values P_{MAX}, P_{DC} and V_{DC} for calculating R_{PM} are those nearest to the zone of slope change in MPPT efficiency curve.

To compute the parameters values, measured efficiency data from inverters were used. As shown in Figure 6, for every value of output power we have several values of efficiency, and therefore various parameters values. To obtain a single value for each parameter, an algorithm was created that uses the values of R_S and R_P (or R_{SM} and R_{PM}) obtained to calculate the efficiency curve for different combinations of these parameters values. These curves are compared with the actual curve and the algorithm chooses the one with a better correlation coefficient.

Model Validation

To validate the efficiency part of the model, we need to measure I_{DC}, V_{DC}, P_{DC}, V_{AC} and P_{AC}, and extract the parameters R_S and R_P using the method described in 3.1. The block diagram of the experimental test setup used to measure is shown in Figure 7. The inverter input current and voltage are measured using channel A of a precision power analyzer, and channel B is used for output values. The records were made in sunny days, taking at least two days of measuring. The computer stores and processes the data.

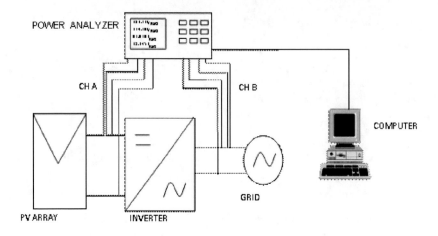

Figure 7. Block diagram of the experimental test setup for the evaluation of the conversion efficiency.

With P_{DC} and P_{AC}, we can calculate η using (3). Then, with I_{DC} and V_{AC} we find the seeds for R_S and R_P with (27) and (29). The extraction algorithm uses all these measured values and also V_{DC}. With the final values for R_S and R_P, we can calculate again η using now (10), and compare both values with two statistics index:

1. Pearson's correlation, that gives the quality of the curve fitting between the calculated efficiency and the measured data.
2. Root mean square error (RMSE), as a measure of standard deviation.

These statistics have been widely used in similar works [20, 41].

The model has been verified with a set of inverters at the Industrial Technical Engineering School of the Polytechnic University of Madrid (ETSIDI – UPM) and in the Laboratory of Photovoltaic Systems at the Research Centre for Energy, Environment and Technology (CIEMAT).

For the inverters at ETSIDI-UPM, the results are shown in Figure 8 – 9 and Table 2.

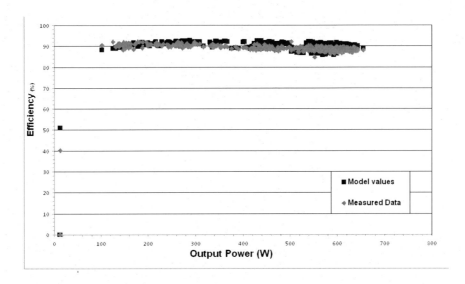

Figure 8. Efficiency of the inverter "A" at EUITI-UPM in terms of output power. Orange diamonds are experimental data, and blue squares are model values.

For comparison, Figures 8 and 9 shows Jantsch et al. model [32] in addition to the proposed model. In this model, efficiency depends on the self-consumption, the load dependent losses, and the output power by the equation:

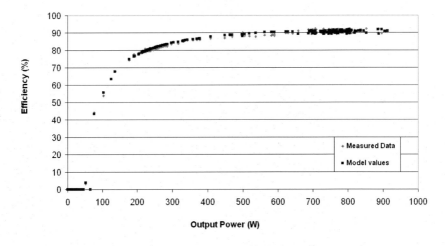

Figure 9. Efficiency of the inverter "B" at EUITI-UPM in terms of output power. Orange diamonds are experimental data, and blue squares are model values.

Table 2. Numerical results for inverters at EUITI-UPM

Inverter	Nominal Power (W)	R_S (Ω)	R_P (Ω)	Pearson's correlation	RMSE (%)
A	850	0.26	4353	0.99888	1.25
B	3000	0.6	1200	0.99958	0.98

$$\eta = \frac{P_{out}}{P_{out} + k_0 + k_1 P_{out} + k_2 P_{out}^2} \tag{30}$$

Were the self-consumption are represented by k_0, the load linear proportional losses by k_1 and the load square proportional losses by k_2.

As can be seen, there are a good fit between the real data and the models. However, Jantsch model for inverter A shows a RMSE of 2.32%, almost twice that the proposed model. For inverter B, Eq. 30 shows a better fit than the new model (Eq. 10), but for this good fit it has been necessary to adjust the k_0 parameter from the measured one (0.16%) to an out of range 1.33%, so that the values cannot be considered valid. Using the correct value for k_0, the RMSE reaches 5.6%. As demonstrated, the fitting between the proposed model and measured efficiency is more accurate than with other models.

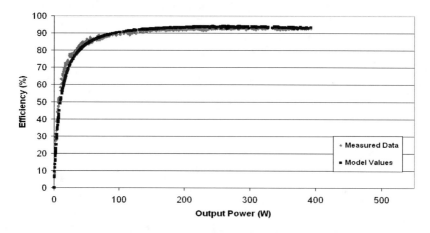

Figure 10. Efficiency of the inverter "C" at CIEMAT in terms of output power.

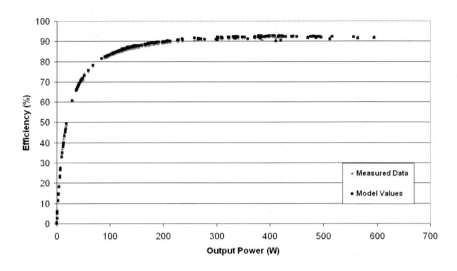

Figure 11. Efficiency of the inverter "D" at CIEMAT in terms of output power.

For the inverters at CIEMAT, Table 3 is a summary of the parameters and statistics, and the graphical results are shown in Figure 10 – 12.

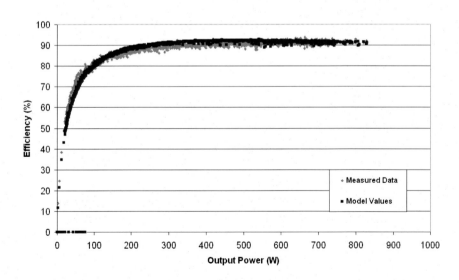

Figure 12. Efficiency of the inverter "E" at CIEMAT in terms of output power.

Table 3. Numerical results for inverters at CIEMAT

Inverter	Nominal Power (W)	R_S (Ω)	R_P (Ω)	Pearson's correlation	RMSE (%)
C	700	2.6	5498	0.99426	1.57
D	850	1.2	3004	0.99986	0.77
E	2000	2.8	2224	0.99702	1.39

To validate the efficiency in tracking the maximum power point, you need to measure V_{DC}, P_{DC}, G and T (the latter parameters are necessary to calculate I_{MAX}, V_{MAX} and P_{MAX} using eq. shows in [31]) and extract the parameters R_{SM} and R_{PM} as described in 3.1. The measuring system can be the same to the one shown in Figure 7. The procedure is virtually identical to that used to validate the efficiency.

Testing was performed using one of the inverters installed at ETSIDI / UPM. The parameters and errors indicated in Table 4 were obtained, and the graph of Figure 13 summarizes the results.

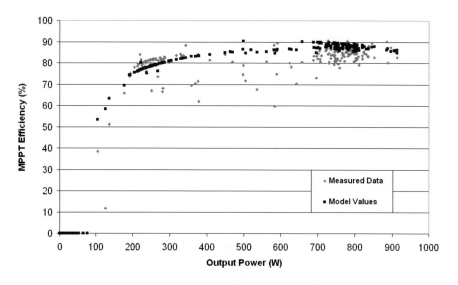

Figure 13. MPPT efficiency for 'B' inverter. Orange diamonds are experimental data, and blue squares are model values.

Table 4. Parameters values and statistical error from 'B' inverter

Inverter	Nominal Power (W)	R_{SM} (Ω)	R_P (Ω)	Pearson's correlation	RMSE (%)
B	3000	1.1	267	0.9966	2.7

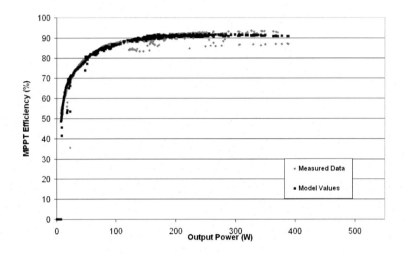

Figure 14. MPPT efficiency for 'C' inverter.

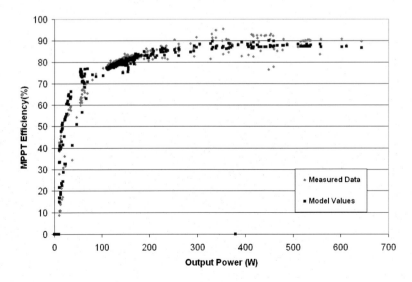

Figure 15. MPPT efficiency for 'D' inverter.

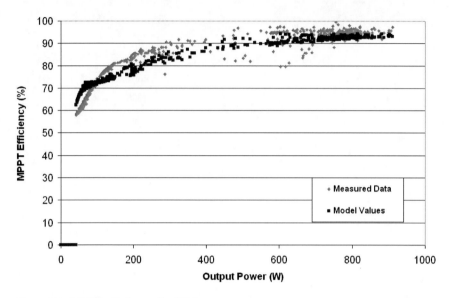

Figure 16. MPPT efficiency for 'E' inverter.

In addition, measurements were made with the 3 inverters installed at CIEMAT. Figures 14-16 show the graphs of the modeled and real MPPT efficiency of such inverters, showing the small dispersion between the values obtained by modeling the inverters and the curves obtained with actual measurements.

The values of the dissipative parameters are shown in Table 5.

In this table we can see that for these inverters, the R_{SM} resistance ranges between 0.18 and 5.4 ohms and R_{PM} varies between 423 and 1802 ohms.

Table 5. R_{SM} and R_{PM} parameters for inverters at CIEMAT

Invertir	Nominal Power (W)	R_{SM} (Ω)	R_{PM} (Ω)	Pearson's correlation	RMSE (%)
C	700	5.4	1802	0.9942	1.98
D	850	2.3	423	0.9981	1.69
E	2000	0.18	673	0.9983	2.20

In terms of error, the modeling of the MPPT efficiency exhibits errors greater than those obtained in the modeling of the efficiency, but largely coincides with what has been achieved experimentally, the error reaching the value of 3.4 % in the worst case. Compared with the model in [31], which uses a constant value for MPPT efficiency, the error is even greater, in the range of 15%.

With this in mind, we can conclude that, although not as closely as for efficiency, the proposed model for efficiency on maximum power point tracking of a PV inverter correctly simulates this characteristic, and it can be implemented in a straightforward manner to be a part of this model for energy performance of inverters.

RESULTS AND APPLICATIONS

The accuracy of the model presented here can be verified by comparing the values of a simulation with real data. Figure 17 shows the comparison between simulated and actual values for input and output voltages and currents, and for input and output power of an inverter. The graph shows in purple color the curves corresponding to the values of the simulation, and in green color, the curves with the measured data. The upper part of the graph shows the DC voltage; we can see that the greatest differences are for low power values. In the middle we have the curves corresponding to the DC and AC currents, with a good fit between them. The lower part shows the input and output power of the inverter; being a losses model, the best fit is achieved in the power values.

For the power curves, it should be noted that the value of the error does not exceed in any case 1 %, with the value in almost all the simulation below 0.5 %. The largest discrepancies occur for low power values, because in that area, it is more difficult to achieve a good fit of the model.

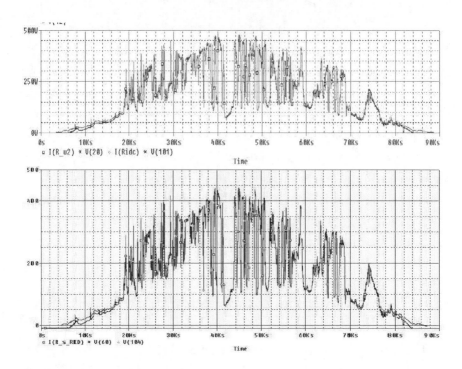

Figure 17. Input and output power (in Watts) on an inverter over a day.

The model can be simulated with any simulator that enables you to enter the corresponding equations. In our case, we have chosen the PSPICE simulator to perform the application examples. The dissipative parameters can be easily simulated by resistors, the ideal inverter is made with dependent sources and disconnection in abnormal utility conditions is done with controlled switches. The simulation of the reconnection is performed by a transmission line with a delay equal to T_R. In this way the inverter does not energize the utility line in the time required after the utility service voltage and frequency have recovered to within the specified ranges.

Figure 18 presents a simulation of the implemented model made in PSPICE. This is a study of the energy generated in a day. The top graph represents the incident radiation, and the bottom one is the energy produced throughout the day by the photovoltaic system. It is noted that its total production is approximately 3.7 kWh.

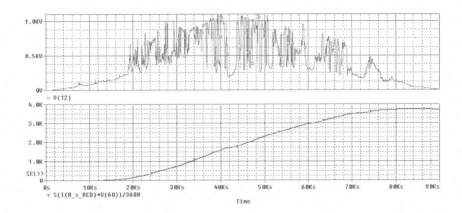

Figure 18. Energy produced (in Wh) in the course of a day. In the upper graph, 1 V represents 1 W/m².

This model has been integrated into a simulation package, which is currently used for educational purposes in the courses offered by our working group [42]. The package allows you to simulate both, stand alone and grid connected PV systems.

Another possible use of the model is the diagnosis of PV systems and fault detection by simulation and comparison with real data. In the process of fault diagnosis in a photovoltaic system, it is common that the true cause of the problem is masked by the symptoms of the fault. Using the simulation as a diagnosis tool, you can minimize the number of parameters to measure and use it as a resource to calculate, statistically, the most probable cause of fault.

Figure 19 represents the energy generated in two days using a photovoltaic system. It can be seen that the energy generated on the first day is virtually identical to the real data curve (in green) and the simulated one (in purple), while the second day both of these graphs differ in approximately 1 kWh, being the real the smaller. This suggests that the system has had a problem and has remained disconnected from the utility.

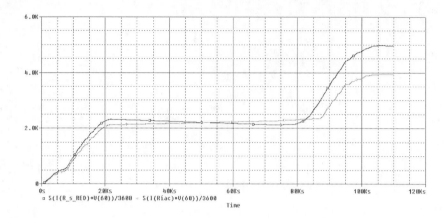

Figure 19. Energy produced (in Wh) over two days.

Figure 20 shows the current output of the inverter. As seen, there has been a disconnection for approximately an hour and a half after that the radiation exceeds the start threshold (see real data curve in green). Subsequent studies attribute the shutdown to a cooling failure, which was revised in this inverter.

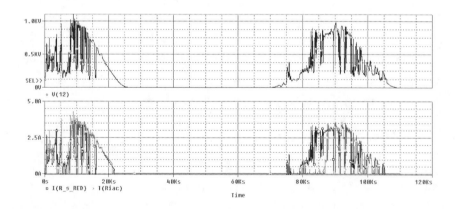

Figure 20. Radiation (top graph) and output current (lower graph) for the inverter under study. In the upper graph, 1 V represents 1 W/m^2.

CONCLUSION

A model that allows simulation of grid-connected PV inverters has been presented. The model is composed of electrical items that simulate the behavior of the electronic system. The model proposes that efficiency, MPPT, and standby consumption are simulated using resistive elements, as shown by the equations [10, 13 and 21]. The other characteristics are simulated by simple circuits.

The proposed model can be used to characterize any inverter, regardless of their topology and the MPPT control method, against the other models presented in this chapter. In addition, it has less parameters than other models. Compared with a similar model, if a correct adjustment of parameter is made, the proposed one presents half the value of error in the efficiency.

A systematic methodology that can be applied to grid-connected PV inverters is proposed. This methodology is useful for the extraction of the parameters included in the model. The proposed method enables the characterization and simulation of any grid-connected inverter. Using this methodology, several inverters have been characterized and it has been proved that the model reflects with great fidelity the actual behavior, with errors less than 3.5%.

The model can be implemented in most of the software packages used in computer simulation, such as SIMULINK, TRNSYS, SPICE, etc. We present some examples of simulations with PSPICE, which check the accuracy of the model when compared with actual data.

The chapter also suggests applications of this model in education and maintenance of PV plants. It can also be a useful tool in the design and implementation of photovoltaic projects.

REFERENCES

[1]   Canmet ENERGY. *Clean energy project analysis: Retscreen® engineering & cases textbook. Introduction to clean energy project analysis.* 3th ed. Canada: Minister of Natural Resources Canada; 2005. Available from: http://www.retscreen.net/

[2]   Mermoud, A. PVSYST: a user-friendly software for PV-systems simulation. In: *Proc. of the 12th European PV Conference,* Amsterdam; 1994.

[3]   Turcotte D. Photovoltaic hybrid system sizing and simulation tools: Status and needs. In: *Proc. of PV Horizon Workshop.* Montreal; 2001.

[4]   Givler T, Lilienthal P. Using HOMER Software, NREL's Micropower Optimization Model, to explore the role of Gen-sets in Small Solar Power Systems. *Technical Report NREL/TP-710-36774.* Available from: http://www.nrel.gov/docs/fy05osti/36774.pdf Last access: 04/18/2013.

[5]   *PV-Design Pro - One of the solar energy programs on the Solar Design Studio v6.0.* Available from: http://www.maui solarsoftware.com/. Last access: 04/18/2013.

[6]   Perpiñán O. solaR: Solar Radiation and Photovoltaic Systems with R. *J. of statistical softw.* 2012:50(9):1-32.

[7]   Castro M, Delgado A, Argul FJ, Colmenar A, Yeves F, Peire J. Grid-connected PV buildings: analysis of future scenarios with an example of Southern Spain. *Sol. Energy* 2005:79(1):86-95.

[8]   Calais M, Hinz H. A ripple-based maximum power point tracking algorithm for a single-phase, grid-connected photovoltaic system. *Sol. Energy* 1998:63(5):277-82.

[9]   Schulz D, Hanitsch R. Short time simulation of solar inverters. In: *Proc. of 16th European PV Conference,* Glasgow; 2000.p. 2430-3.

[10] Trujillo CL, Velasco D, Figueres E, Garcerá G, Ortega R. Modeling and control of a push–pull converter for photovoltaic microinverters operating in island mode. *Appl. Energy* 2011: 88:2824–34.

[11] Sharaf AM, Ozkop E, Altas IH. A hybrid photovoltaic PV array-battery powered EV-PMDC drive scheme. In: *Proc of IEEE Electr. Power Conference, Montreal;* 2007.p. 37-43.

[12] Ortjohann E, Voges B, Voss J. Measurement and modelling of a grid connected photovoltaic inverter. In: *Proc. of the 11th European PV Conference, Montreux;* 1992.p. 1379-81.

[13] Bettenwort G, Bendfeld J, Drilling C, Ortjohann E, Rump S, Voß J. Model for evaluating MPP methods for grid-connected PV plants. In: *Proc. of 16th European PV Conference,* Glasgow; 2000.p.2578-81.

[14] Orduz R. Analytical study and evaluation results of power optimizers for distributed power conditioning in photovoltaic arrays. *Progress in Photovolt.,* 2013:21(3):359–73.

[15] Molina M, Mercado P. Modeling and Control of Grid-connected Photovoltaic Energy Conversion System used as a Dispersed Generator. *IEEE Transmission and Distribution Conference and Exposition*, Bogotá; 2008.p.1-8.

[16] Crastan V. Improvemente of solar inverters for grid connected PV-systems. In: *Proc of 14th European PV Conference, Barcelona;* 1997.p. 2237-8.

[17] Tsai HL, Tu CS, Su YJ. Development of generalized photovoltaic model using MATLAB/SIMULINK. In: *Proc. of Int. Conference on Electr. Eng. and Appl.,* San Francisco; 2008.

[18] Villalva MG, Gazoli JR, Filho ER. Comprehensive approach to modeling and simulation of photovoltaic arrays. *IEEE Trans. on Power Electr.* 2009;24(5):1198-208

[19] Sheriff F, Turcotte D, Ross M. PVTOOLBOX: A Comprehensive Set of PV System Components for the Matlab/Simulink

Environment. In: *Proc. of Conference of the Sol. Energy Soc. of Canada,* Ontario; 2003.

[20] Guasch D. *Modelado y análisis de sistemas fotovoltaicos [Modeling and analysis of photovoltaic systems].* Doctoral Thesis. Politechnic University of Cataluña:2003. Spanish.

[21] Silvestre S, Castañer L, Guasch D. Herramientas de Simulación para Sistemas Fotovoltaicos en Ingeniería [*Simulation Tools for Photovoltaic Systems in Engineering*]. *Formación Universitaria,* 2008:1(1): 13-8. Spanish.

[22] Smith GA, Onions PA, Infield DG. Predicting islanding operation of grid connected PV inverters. *IEE Proc. on Electr. Power Appl.,* 2000: 147(1):1- 6.

[23] Simmons A, Infield D. Current waveform quality from grid-connected PV inverters and its dependence on operating conditions. *Progress in Photovolt.,* 2000:8(8):411–20.

[24] Ichinokura O, Maeda M, Takahashi H, Murakami K. Fault simulation of orthogonal core type DC-AC convert for PV power system. In: *Proc. of the 11th European PV Conference,* Montreux; 1992.p. 1182-5.

[25] Schilla T, Hill R, Pearsall N, Forbes I, Bucher G. Development, simulation and evaluation of digital maximum power point tracker models using Pvnetsim. In: *Proc. of 16th European PV Conference,* Glasgow; 2000.p. 2482-5.

[26] Aguilar JD, Nofuentes G, Marín J, Hernández JC, Muñoz FJ, Guzmán E. *Estudio de la célula solar con ayuda de simulador PSPICE® y de medidas de módulos fotovoltaicos de silicio cristalino a sol real* [Study of the solar cell with the help of PSPICE® simulator and measurements of photovoltaic modules of crystalline silicon to real sun]. In: *Proc of 7th Conference on Technol. Appl. to Electr. Teach* (TAEE 06), Madrid, 1996. Spanish.

[27] Pongratananukul N, Kasparis T. Tool for automated simulation of solar arrays using general-purpose simulators. In: *Proc of 9th*

*EEE Workshop on Comput. in Power Electr. (*COMPEL), Illinois, 2004.p. 10-4.

[28] Roche D, Outhred H, Kaye RJ. Analysis and Control of Mismatch Power Loss in Photovoltaic Arrays. *Progress in Photovolt.*, 1995:3(2):115–27.

[29] Nofuentes G, Hernández JC, Almonacid G, Abril JM. Analysis of the possibilities to omit the blocking and bypass diodes using a standard circuit simulator. In: *Proc. of the 14th European PV Conference,* Barcelona; 1997.p. 1094-6.

[30] Silvestre S, Boronat A, Chouder A. Study of bypass diodes configuration on PV modules. *Appl. Energy* 2009:86:1632–40.

[31] Castañer L, Silvestre S. *Modelling photovoltaic systems using Pspice.* Wiley; 2002.

[32] Jantsch M, Schmidt H, Schmid J. Results of the concerted action on power conditioning and control. In: *Proc. of the 11th European PV Conference,* Montreux; 1992.p. 1589-93.

[33] Chivelet NM, Chenlo-Romero F, Alonso-Garcia MC. Modelado y fiabilidad de inversores para instalaciones fotovoltaicas autónomas a partir de medidas con cargas resistivas y reactivas [Modeling and reliability of inverters for autonomous photovoltaic installations from measurements with resistive and reactive loads]. In: *Proc of 7th Iberian Conference on Sol. Energy,* Córdoba; 1994. Spanish.

[34] Chivelet NM, Chenlo-Romero F, Alonso-Garcia MC. Analysis and modelling of DC/AC inverters with resistive and reactive loads for stand-alone PV systems. In: *Proc. of the 12th European PV Conference,* Amsterdam; 1994.

[35] Koizumi H, Kaito T, Noda Y, Kurokawa K, Hamada M, Bo L. Dynamic response of maximum power point tracking function for irradiance and temperature fluctuation in commercial PV inverters. In: *Proc. of the 17th European PV Conference,* Munich; 2001.

[36] Al-Almoudi A, Zhang L. Maximum power point tracking using a neural network model for grid-connected PV systems. In: Proc. of the 2nd World Conference and Exhib. on PV Sol. *Energy Convers., Viena*; 1998. p. 2049 – 53.

[37] Cendagorta M, Galbas R, Monzón MR, Dobón F, Pérez F, García B et al. Differential MPP controlling. In: *Proc. of the 2nd World Conference and Exhib. on PV Sol. Energy Convers.,* Viena; 1998. p. 2112 – 4.

[38] Castañer L, Aloy R, Carles D. Photovoltaic system simulation using a standard electronic circuit simulator. *Progress in Photovol.,* 1995: 3(2):239 – 52.

[39] Stevens J, Bonn R, Ginn J, Gonzalez S. Development and Testing of an Approach to Anti-Islanding in Utility-Interconnected Photovoltaic Systems. *Technical Report SAND 2000-1939. Sandia National Laboratories.* Alburquerque, 2000.

[40] Davila-Gomez L, Castro-Gil M, Colmenar-Santos A, Peire J. Measuring and static modelling of grid connected PV inverters. In: *Proc. of the ISES Sol. World Congress,* Göteborg; 2003.

[41] Chouder A, Silvestre S, Sadaoui N, Rahmani L. Modelling and simulation of a grid connected PV system based on the evaluation of main PV module parameters. Simul. *Modelling Practice and Theory* 2012:20(1):46–58.

[42] Muñoz J, Díaz P. A virtual photovoltaic power systems laboratory. In: *Proc. of the 1st IEEE Education Engineering Conference* (EDUCON), Madrid; 2010.p.1737-40.

In: Solar Power Technology   ISBN: 978-1-53614-204-4
Editors: A. Colmenar-Santos et al. © 2019 Nova Science Publishers, Inc.

Chapter 4

PV POTENTIAL IN SHOPPING CENTERS

Severo Campíñez-Romero[1,], África López-Rey[1] and Jorge-Juan Blanes-Peiró[2]*

[1]Departamento de Ingeniería Eléctrica, Electrónica, Control, Telemática y Química Aplicada a la Ingeniería, Universidad Nacional de Educación a Distancia (UNED), Madrid, Spain
[2]Universidad de León, León, Spain

ABSTRACT

The solar photovoltaic is a renewable energy source which allows nowadays, in many places, the generation of electricity at a cost comparable with the conventional thermal generation methods. However, ending 2014, with a worldwide power capacity installed of 177 GW, the integration level in the electricity generation mix was still far from the targets set up for this technology in the global warming mitigation strategies. Technical, financial and regulatory barriers slow down a massive penetration of photovoltaic generation facilities, hence practical and innovative measures are necessary to facilitate a wider deployment. The building integration of solar photovoltaic facilities offers an efficient

* Corresponding Author Email: s.campinez.romero@gmail.com.

solution because the electricity is generated near the consumption point and gives a new economic value to roofs and facades. In this chapter, we develop a scientist methodology for determining the potential in shopping centers, to establish that, in terms of photovoltaic integration, these present important competitive advantages over other alternatives, i.e., the residential buildings. The power capacity potential obtained making use of this methodology in the countries selected is 16,8 GW, that means 10% of the worldwide capacity installed at the end of 2014, with a yearly electricity generation of 22,7 TWh equivalent to 14% of the worldwide generation in 2014.

Keywords: solar energy, building integration, grid integration, shopping centres

NOMENCLATURE

General

A_{SP}	Area of the solar panel
A_{PV}	PV area
CR	*Covering ratio*
DOE	US Department of Energy
DSO	Distribution System Operator
EU28	European Union 28 countries
GLA	Gross Leverage Area
GHG	Greenhouse gases
G_T	Yearly solar irradiation (insolation) in kWh/m^2/year incident on an optimally-tilted solar panel
P_P	PV Peak power
P_{SP}	Maximum power of the solar panel at STC conditions
PR	Performance ratio of the PV facility
PV	Photovoltaic electricity source or facility
RES	Renewable Energy Source
SC	Shopping Centre
US	United States of America

Greeks

α_s	Solar altitude angle
β	Panel slope. It is the angle between the solar panel surface and the horizontal.
γ_s	Azimuth solar angle
ε	Solar panel efficiency
φ	Latitude angle

INTRODUCTION

Global warming experienced in the earth in the last decades is a reality with proved grave consequences. With a high degree of probability, the key reason behind this global warming is the anthropogenic greenhouse gasses (GHG hereinafter) emissions, fundamentally caused by the global population growth rate and the economic development since the beginning of the industrial era (Intergovernmental Panel on Climate Change, 2014).

To face this situation, the international community has adopted the commitment to limit the earth global warming to 2°C above pre-industrial levels (European Climate Foundation, 2010; The White House. President of the US, 2013). The fulfilment of this commitment requires of global and important measures sustained over time to achieve a significant reduction of the GHG emissions. Many governments and international institutions have established future scenarios and designed strategies to reach this achievement that include necessarily interventions to mitigate the present emission levels.

According to the majority of the analysis carried out in this field, one of the main sources of GHG emissions is the power sector; therefore, important efforts should be done here to contribute as it should be, towards the achievement of the mitigation targets. The use of renewable energy sources (RES hereinafter) in the electricity generation is the base to support this contribution. Presently it has been reached a significant share of penetration of RES in the electric generation mix; but this rate should be increased massively to attain the transition from the present electricity generation scheme, fundamentally based on the use of hydrocarbons (Intergovernmental Panel on Climate Change, 2011), to a near-zero emissions power sector

Among the RES, the solar photovoltaic source (PV hereinafter) has an essential characteristic that favors its massive penetration: the dispersion degree. The solar radiation is received worldwide with an intensity level that, depending on the specific location latitude, makes possible the production of electricity almost everywhere. Besides, the PV technology presents nowadays a technical and economical maturity that allows high efficiency generation at such a low cost comparable with the traditional generation based on conventional thermal methods (Colmenar-Santos et al., 2012; Feldman et al., 2014; Hernández-Moro and Martínez-Duart, 2013; Lazard, 2014; Ondraczek et al., 2015; Ossenbrink et al., 2013; Philipps et al., 2014).

The integration of a massive share of RES in the electric power grid implies technical barriers and problems with an impact on the operation and stability of the whole system. The facilities must implement specific systems to favor the integration of the electricity generated, including eventually elements to storage the surplus generation to be used in deficit times (Barth et al., 2014; Luthander et al., 2015; Mai et al., 2012). This is a handicap for small facilities in residential buildings or single family houses, even for small commercial buildings, because all of these auxiliary systems can turn the facility more complex and with needs of professional assistance and likely their users will not have an adequate technical profile and will have to contract a utility company to do it.

In the previous technical literature, several studies have been carried out to determine roof area availability or solar PV potential in buildings. Most of them are focused only on residential buildings:

(Karteris et al., 2013) analyzed the suitable roof areas in multifamily buildings in cities based on statistical calculation obtaining that the solar roof factor of utilization in this type of residential buildings is only between 25% and 50%. (Li et al., 2015) investigated the annual solar yield per floor space in urban residential buildings at different levels of site density. (Ordóñez et al., 2010) developed a methodology to determine the solar PV generation potential on residential rooftops using statistical data to characterize the building stock.

Other previous research did not differentiate the use of the buildings under analysis and, therefore or do not detail the PV potential of every of them: (Defaix et al., 2012) estimated the technical potential in integrated PV for residential and non-residential buildings in the EU-27 countries, (Izquierdo et al., 2008) defined a methodology, based on statistics available data and GIS maps, for estimating roof surface area for large-scale PV potential evaluations in all type of existing buildings and applied this methodology to Spain to obtain an estimate of energy generation, (Peng and Lu, 2013) carried out an investigation about the PV potential of hotels and commercial buildings in the city of Hong Kong. (Pillai and Banerjee, 2007) developed a methodology to for potential estimation of solar water heating that applied to a synthetic area of 2 sq. km in India considering residential houses, hospital, nursing homes and hotels. (Schallenberg-Rodriguez, 2013) investigated about the solar PV potential on roofs for several type of buildings in regions and islands and applied their research for obtaining an estimate of the PV potential in the Canary Islands, (Singh and Banerjee, 2015) carried out a methodology for estimating solar photovoltaic potential in a city for all type of buildings and (Wiginton et al., 2010) defined a method to estimate total rooftop PV potential based in GIS and object-specific image recognition city of Ontario (Canada).

Generally, the pre-existing literature has considered any type of buildings with similar features to host PV facilities independently of its use. However, from a constructive point of view, the availability of residential building roofs is generally lower to commercial building ones.

All of the aforementioned features, as well as other analyzed in more detail below, point to shopping centers as best-fitting buildings for the purpose of installing PV facilities and therefore they should be the target of a detailed assessment about their PV potential.

As a consequence, the likelihood of taking advantage more efficiently of the roof space and the better technical adaptability for the integration of PV generation contributes to maximize the effect in the building energy performance because it is possible to achieve a larger level of self-generation.

This chapter is aimed firstly at analyzing the particular advantages of shopping center buildings to install PV facilities, including technical comparatives versus the residential buildings that represent the predominant type widely used to estimate the PV potential. Secondly, the article focuses the research on determining the shopping centers potential in terms of power capacity and electricity generation as well as the possible contribution to achieve the penetration targets set up for this technology. The research has been developed for a large area including all the countries in the 2014 PV power capacity top ten list.

Below in chapter 2 we review the world present situation in terms of photovoltaic penetration and will be determined the reference targets used in this article. In chapter 3, we specify the reasons for considering shopping centers as buildings with important advantages for PV facilities. In chapter 4 we define the methodology developed to obtain the power capacity and generation potential in shopping centers located in different countries and in the chapter 5 we carry out the determination of this potential. In the chapter 6 the methodology is applied to a real shopping center located in Madrid (Spain) to obtain the capacity and share potential. Finally, in the chapter 7 the conclusions of the results obtained in the study are presented.

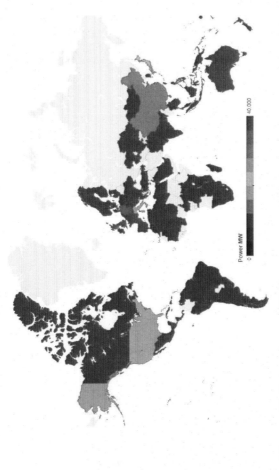

Source: (International Renewable Energy Agency, 2015).

Figure 1. Distribution of the PV installed power per country at the end of 2014.

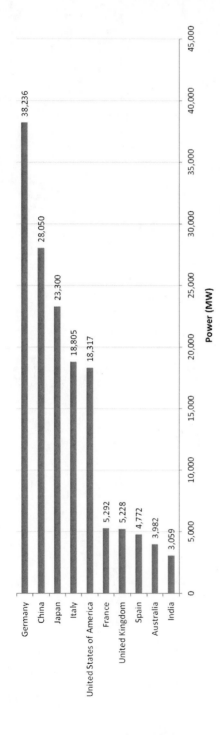

Source: (International Renewable Energy Agency, 2015).

Figure 2. Top ten countries by PV installed power.

PRESENT SITUATION AND TARGETS
PRESENT WORLD SITUATION

Ending 2014, the world PV power capacity installed was 177 GW. As shown in Source: (International Renewable Energy Agency, 2015).

Figure 1 the distribution of the capacity presents a high concentration degree, so that the top ten countries host 149 GW, what means 84% of total (Figure 2).

The countries included in this article are those in the top ten list: the 28 countries members of the European Union (EU28 hereinafter) including Germany, Italy, France, the United Kingdom and Spain; China, Japan, the United States of America (US hereinafter), Australia and India.

PHOTOVOLTAIC INTEGRATION TARGETS

Several international governmental and non-governmental organizations have set up scenarios and paths focused on the power sector aimed to reduce its GHG emissions. Among all of that, those considered in this article, with a deployment period up to 2050, are the following:

- In the US, the targets set up in the SunShot Vision Study (National Renewable Energy Laboratory, 2012), which was developed under the US Department of Energy (DOE) (The US Department of Energy, 2015) SunShot Initiative. It is focused on the solar photovoltaic renewable source to be competitive with the traditional generations forms before 2020, favoring its integration and contributing to the achievement of the GHG emissions mitigation and reduction plans adopted by the Obama Administration in 2009 and 2013 (The White House. President of the US, 2013) and to the Clean Power Plan (U.S. Environmental Protection Agency, 2015), which has been recently implemented in 2015 to reduce the carbon emissions in the electricity generation facilities.

- In the 28 countries members of the EU28, the targets established in the Roadmap 2050 (European Climate Foundation, 2010), that analyses and sets up the paths to achieve the European commitment to reach in 2050 GHG emissions below 80% of 1990 levels.

- In the rest of countries included in this article: Australia, China and India, the targets are those set up by the International Energy Agency in the high-renewables (IEA hi-Ren) scenario of the Energy Technology Perspectives 2014 report (ETP2014) (International Energy Agency, 2014a), where is stablished the limitation for the global warming in 2°C above pre-industrial levels with a high share of RES penetration.

Figure 3 and Figure 4 illustrate a comparative between the capacity and share figures of PV ending 2014 and the targets foreseen for 2050 for the countries included in the study. It is noticeable that in all the cases, the increase necessary for both the capacity and the share to fulfil the targets. According to the IEA hi-Ren, the worldwide power capacity must go from the present 177 GW to 4,674 GW, what means a sustained yearly increasing of 10% up to 2050; the share will go approximately from 1% to 16%. The region presently closest to achieve its targets is EU28, with 86.8 GW of power capacity, mainly in Germany (38.2 GW or 44% of total) and a share of 3.1%. EU28 will need a sustained yearly growing of 6% for power capacity and 5% for share to reach the Roadmap 2050 targets. To achieve the objectives set up in the SunShot Vision Study, US will need a sustained annually increasing for both the capacity power and the share of 10%, to go from the present 18.3 MW up to 632 MW and from 0.6% to 19% respectively. Finally, in China and India the yearly growth should be of 12% y 16% respectively to get the targets established in the IEA hi-Ren and 11% y 12% to achieve the share estimates.

Present Integration in Shopping Centres

Nowadays, the PV capacity integrated in shopping centers worldwide is irrelevant. In example, in the US, ending 2014, there was 325 MW (Solar Energy Industries Association, 2014), barely 1,5% of total country PV capacity.

WHY THE SHOPPING CENTRES?

There are diverse types of buildings such as schools, hospitals or industrial and commercial buildings that present particularities and different characteristics that advices to analyze each of them with different approaches in order to estimate their PV potential. In this article we have focused the research in developing a new methodology applicable to the shopping centers, using the residential buildings as a benchmark for highlighting the advantages for the implementation of PV integrated facilities versus the predominant building type.

These advantages of shopping centers are detailed, justified and compared along this chapter.

CONSTRUCTION CHARACTERISTICS

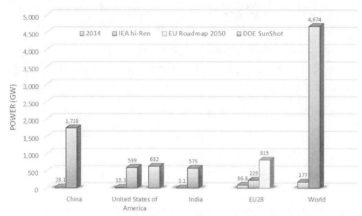

Sources: (European Climate Foundation, 2010; Feldman et al., 2015; International Energy Agency, 2014a; International Renewable Energy Agency, 2015; Johnston et al., 2015; National Renewable Energy Laboratory, 2012; REN21, 2015; RTE Réseau de transport d'électricité, 2015; Yamada and Ikki, 2015).

Figure 3. Present PV power installed and targets for 2050.

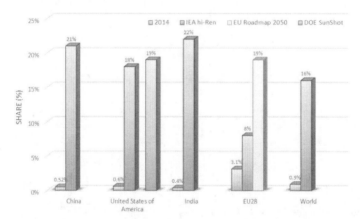

Sources: (Arbeitsgemeinschaft Energiebilanzen, 2014; Australian Energy Regulator, 2014; British Petroleum, 2015; European Climate Foundation, 2010; Feldman et al., 2015; International Energy Agency, 2014a; International Renewable Energy Agency, 2015; Johnston et al., 2015; National Renewable Energy Laboratory, 2012; Red Eléctrica de España S.A., 2015; RTE Réseau de transport d'électricité, 2015; U.S. Energy Information Administration, 2015; UK Government. Department of Energy & Climate Change, 2015; Yamada and Ikki, 2015).

Figure 4. Present PV share and targets for 2050.

Shopping centers present construction features that favor the installation of PV in the roofs. As it is shown in (Denholm, 2008) the availability rate for residential buildings is between 22% and 27% and some other studies set up this rate in the 25% - 50% range for high population density cities where predominate multifamily buildings (Karteris et al., 2013). On the other hand, the availability rate for commercial buildings can be placed between 60% and 65% (Denholm, 2008). This means that, in similar conditions, it will be possible the installation of double or triple capacity in shopping centers roofs than in residential buildings.

SELF-CONSUMPTION AND CONSUMPTION PROFILE

In terms of transport and distribution of energy, the most efficient place to generate electricity is nearest the consumption point. In this matter, the shopping centers offer also relevant advantages over other building types. On one hand because shopping centers are major electricity consumers and the generation can be mostly used for self-consumption, and on the other hand because some studies show that their consumption patterns match the PV generation profile better than any other type of buildings (International Energy Agency, 2014b).

Grounded PV facilities sited near high-consumption points present an important handicap, because with the purpose of being near the consumer they will have to be located in urban or industrial areas where the land is highly valued and the rent will penalize the PV operation costs. However, the shopping centers, by nature, are usually located near o inside cities or populated and industrial areas and the electricity surplus could be used to supply the local consumption through the distribution networks with few electric losses.

Last but not less important is the fact that, the self-generation of electricity will reduce the energy dependency of the building, contributing to increase the performance of the shopping centers in terms of efficiency.

Integration Challenges

The electricity generated in PV facilities has a non-manageable character; it means that it is not possible to control instantaneously the generation level except to reduce it. This characteristic makes difficult a massive penetration therefore the implementation of specific systems and technical measures are necessary for an adequate management (Barth et al., 2014).

Some of these measures must be implemented by the distributor system operators (DSO) because their distribution networks were initially designed for a single direction energy flow, and now they must allow the feeding of energy from the consumption points (Stetz et al., 2014). Others will have to be implemented directly in the generation facilities, mainly the following ones:

- Short-term generation forecast.
- Storage of surplus generation to be used in deficit periods.
- Ancillary services for voltage and reactive energy control.
- Demand response.
- Generation curtailment.

Part of them may be implemented in the facility equipment; in example to carry out the voltage and reactive energy control. However, others, as the short-term generation forecast or the demand response will require of regular attention and professional management by the prosumer[1].

---

[1] A prosumer is a consumer and producer of electricity.

In this aspect, shopping centers also present advantages over residential buildings, because they usually have on site industrial technical equipment, as energy supply (electricity, vacuum, fuel), heating and air conditioning systems, security, energy management, etc. and even electricity generation facilities for self-consumption. These systems already need a professional operation and maintenance service (in-house or outsourced) that could include easily the new tasks necessary for the PV management.

DETERMINATION OF THE SHOPPING CENTRES PV POTENTIAL

A novel methodology, illustrated in the Figure 5, has been developed to estimate the PV potential in shopping centers. This new methodology, detailed along this chapter, has been specially designed to be applied in shopping centers. We have used statistical raw data about the characteristics of the buildings to calculate the available roof space in shopping centers, and by means of the definition and determination of reduction ratios, we have finally obtained the PV potential, in terms of power capacity and capacity of electricity generation in a wide world area corresponding with the countries in the top ten list of accumulated PV power in 2014.

CALCULATION OF THE PV ROOF SPACE

The first step in the application of the methodology is determining the PV roof space. It is defined in this article as the shopping center roof surface without physical, technical or shading limitations for placing solar panels on it. The first step to calculate the PV roof space is the determination of the available area in shopping centers. The parameter chosen here to quantify this area is the gross leasable area (GLA hereinafter). GLA is widely used in the real state sector and indicates the total floor area designed for tenant occupancy and exclusive use, including any basements, mezzanines, or upper floors. Therefore, using GLA is a way to classify commercial buildings giving a value of the amount of space available for renting.

Table 1 shows the present GLA values for the countries considered in this article. As it can be seen, the total GLA value is about 845 million of square meters, with a special presence in the US housing 72% of the total.

Although Russia allocates presently around 516 shopping centers representing about 17 million of square meters of GLA, has not been included in this study because the lack of both national regulations to promote PV integration and RES implementation targets.

Not all the shopping centers will be able to host a PV facility because they do not have roof or it is inaccessible. In some cases, on the roofs might be located car parks; and others, mainly in those shopping centers located inside cities or populated areas, may be underground or integrated in buildings with other uses, as hotels or households.

For those centers with roof area limitations due to architectural or functional issues (i.e., parking areas located in the roof), the reduction in the available area will be taken into account by means of the application of the availability rate as it will be defined below in this chapter.

Due to the fact that there is not available statistic data about the constructive characteristics of the shopping centers, it is impossible to obtain an accurate estimate of the quantity and volume in terms of GLA of shopping centers without roof or entirely inaccessible. However, it must be taken in consideration that these shopping centers likely will be integrated in population areas and they will be small and that in the scope of application of this study are included only the shopping centers with a GLA above 5,000 square meters. Nevertheless, it is possible that some shopping centers included in the scope present a total limitation for the installation of PV due to the lack of roof. To consider this limited likelihood, a reduction in the GLA has been considered by the application of a suitability ratio with a conservative value of 0.8.

Once the suitable GLA space is estimated, the next step is calculating the roof space what we have done by means of determining a Roof space/GLA ratio. This ratio will depend on the individual typology and features of every single shopping center, nevertheless to be considered in our method an average value was determined by analyzing the characteristics of a sample of 500 shopping centers spread across the countries included in this study. The analysis consisted in the location and measurement of the individual roof space with Google Earth[TM] (Google Inc., 2013). As an example of the analysis carried out, the Table 2 shows a partial list of the shopping centers included in the sample (the entire list of 500 buildings examined is not included for a matter of space) which average Roof space/GLA ratio is 0.58 equal to the value obtained for the whole sample.

Not all the roof space will be obstacle-free and able to allocate the PV equipment. There might be architectural elements, as skylights or parts with not enough carrying capacity, functional, as heating and air conditioning equipment, and nearby or own constructions shading part of the space. All these elements will reduce the roof space, therefore it is necessary to consider a reduction ratio, named here availability ratio, to obtain the effective PV roof space.

Again, this availability ratio will be particular for every single shopping center according to its typology and constructive characteristics. Several previous studies have determined average values that may be generally applied. A review of them was carried out by Byrne et al. (Byrne et al., 2015). For commercial buildings is widely recommended a value of 0.65 in cold climates and 0.6 in warm climates (Denholm, 2008; Gutschner et al., 2002). Under a conservative approach, the minimum value of 0.6 has been considered in this study.

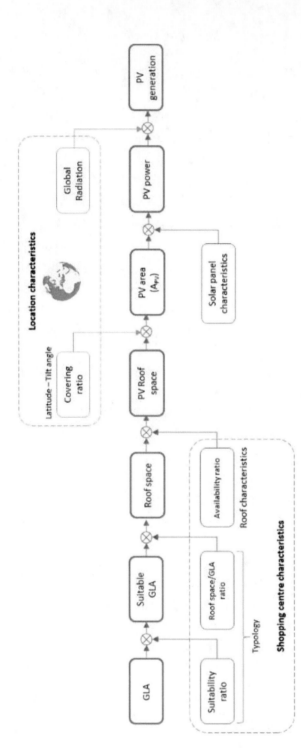

Figure 5. Methodology to estimate PV capacity and generation potential.

Table 1. GLA values for shopping centers with a size bigger than 5,000 m²

Region	Shopping Centre type	GLA range m²	GLA range sq ft	Number of facilities	GLA (m²)	GLA over total (%)
US	Super-Regional Mall	>74,000	>800,000	632	72,449,656	
	Regional Mall	37,000 - 74,000	400,000 - 800,000	608	33,289,047	
	Community Center ("Large Neighborhood Centre")	11,600 - 37,000	125,000 - 400,000	9,642	177,042,603	
	Neighborhood Centre	2,800 - 11,600	30,000 - 125,000	32,422	216,279,091	
	Power Center	23,200 - 55,700	250,000 - 600,000	2,249	91,182,862	
	Lifestyle	13,900 - 46,400	150,000 - 500,000	437	13,590,001	
	Factory Outlet	7,400 - 23,200	50,000 - 400,000	351	7,821,766	
			Total US	46,341	611,655,026	72%
EU28	Traditional	>5,000	>53,800	5,626	118,585,681	
	Retail Park	>5,000	>53,800	2,048	32,479,398	
	Factory Outlet Centre	>5,000	>53,800	162	3,036,653	
			Total EU28	7,836	154,101,732	18%
China	All types	>5,000	>53,800	621	53,200,000	
			Total China	621	53,200,000	6%
Australia	Regional	46,400 - 74,000	500,000 - 800,000	91	7,161,000	
	Sub-regional centers	18,500 - 46,400	200,000 - 500,000	313	6,727,000	
	Neighbourhood/Supermarket-based	1,850 - 18,500	20,000 - 200,000	840	4,340,000	
	Central business district	>5,000	>53,800	123	1,302,000	
			Total Australia	1,367	19,530,000	2%
India	All types	>5,000	>53,800	530	6,500,000	
			Total India	530	6,500,000	1%
			Grand total	56,695	844,986,758	

Sources: US (International Council of Shopping Centers, 2015b), EU28 (Cushman & Wakefield, 2014; International Council of Shopping Centers, 2015a; RegioData Research, 2013), rest of countries (Cushman & Wakefield, 2014).

Table 2. Roof space/GLA ratio calculation in a sample

Country or Region	City	Name	GLA m²	Roof space m²	Roof space/GLA ratio
EU28	Brussels	Woluwe SC	42,677	31,685	0.742
EU28	Madrid	La Gavia	86,356	77,382	0.896
EU28	Prague	Tesco Letnany	125,000	71,472	0.572
EU28	Copenhagen	City Two	63,000	47,501	0.754
EU28	Marseille	Grand Vitrolles	61,111	61,408	1.005
EU28	Paris	Rosny 2	111,600	57,282	0.513
EU28	Berlin	Gropius Passagen	90,000	36,153	0.402
EU28	Dublin	Blanchardstown Centre	100,000	81,054	0.811
EU28	Bucharest	AFI Palace Cotroceni	75,000	75,320	1.004
EU28	Leeds	White Rose SC	63,638	47,017	0.739
EU28	Kyiv	Ocean Plaza	72,200	31,719	0.439
US	Washington	Tysons Corner Centre	220,000	95,648	0.435
US	Los Angeles	Lakewood Centre	200,000	135,632	0.678
US	Massachusetts	South Shore Plaza	201,100	110,405	0.549
China	Beijing	Golden Resources	557,000	61,512	0.110
China	Shangai	Super Brand Mall	121,400	21,837	0.180
India	Mumbai	Neptune Magnet Mall	98,100	20,310	0.207
India	Chennai	Gold Souk Grande Chennai	56,000	15,828	0.283
Australia	Sidney	Macquaire Center	138,500	69,208	0.500
Australia	Robina	Robina Town Center	125,000	99,447	0.796
				Average	**0.580**

Source: (DTZ, 2013) and self-elaboration.

Table 3. PV Roof space calculation

Region	GLA (Ha)	Suitability ratio		Suitable GLA (Ha)	Roof space/GLA ratio		Roof space (Ha)	Availability ratio		PV Roof space (Ha)
US	61,166			48,932	0.58		28,381			17,028
EU28	15,410			12,328			7,150			4,290
China	5,320	x	=	4,256	x	=	2,468	0.6	=	1,481
Australia	1,953	0.8		1,562			906			544
India	650			520			302			181
Total	84,499			67,599			39,207			23,524

As a result of applying the reduction ratios aforementioned, an estimate of PV roof space can be obtained from suitable GLA. The Table 3 shows the detailed calculation for every single country included in the study. The total value reaches 23.5 Ha (over 235 million of square meters).

CALCULATION OF PV POWER CAPACITY AND GENERATION POTENTIALS

Once the PV roof space is determined, the next step is to calculate the power capacity and the generation estimates. With this aim two aspects are essential, on one hand the location of the shopping center, which will determine the solar irradiation, the optimal inclination angle and the distance between panel rows; on the other hand, the technical characteristics of the solar panels and the rest of the equipment forming the PV facility.

Roof PV facilities are typically designed in fixed-mounted execution, so that, they do not change the orientation of the tilt angle along the year. In order to maximize the total[2] irradiation received, the solar panel rows are faced south (in the northern hemisphere and north in the southern hemisphere) with an optimal tilt angle

There are several methods for estimating the global radiation incident on a tilted surface from the horizontal radiation, which is the value typically logged in meteorological observatories, and by means of any of them is possible to obtain the optimal tilt angle in every single day in the year (Hay and Davies, 1980; Hottel and Woertz, 1942; Klucher, 1979; Liu Y.H. and C. Jordan, 1960; Perez et al., 1990; Perez et al., 1987; Perez et al., 1988; Reindl et al., 1990). Nevertheless, for fixed-mounted facilities is widely recommended to use an optimal angle

---

[2] Total radiation is the total sum of the beam, diffuse and reflected or albedo.

value around the site latitude (Gunerhan and Hepbasli, 2007; Lewis, 1987); therefore, it will be the value considered in this study

The tilt angle has a direct effect in the distance between solar panel rows. The classic method to calculate this distance is to consider that the panels will receive the solar radiation without shadows most of the lower solar zenith angle day. The calculation is illustrated in the Figure 6, and the results are shown in the Table 4.

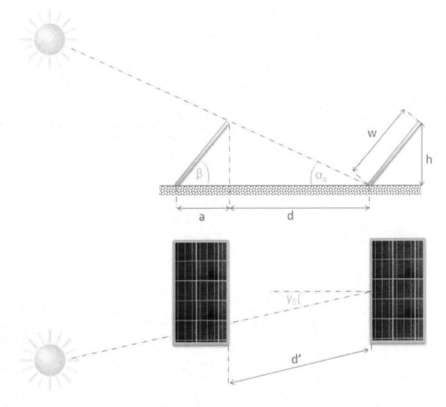

Figure 6. Diagram for the calculation of the distance between solar panels. Azimuth solar panel angle is 0.

Table 4. Calculation of the distance between panels as a function of the site latitude (the slope of the panel is equal to the latitude)

Latitude (φ)	Solar altitude angle (α_s)[a]	Solar azimuth angle (γ_s)[a]	d`	d	a	a + d	Covering ratio w/(a + d)
(deg)	(deg)	(deg)	(cm)	(cm)	(cm)	(cm)	
50	6.66	40.31	524.46	339.3	51.4	390.7	0.20
40	14.23	41.52	202.77	134.4	61.3	195.7	0.41
30	21.60	43.71	101.03	69.8	69.3	139.1	0.58
20	28.63	47.06	50.11	36.7	75.2	111.9	0.72
10	35.15	51.80	19.73	15.5	78.8	94.3	0.85
0	40.91	58.23	0.0	0.0	80.0	80.0	1.00

[a] α_s and γ_s are the value for the latitude angle and the solar azimuth at nine o'clock on the 21st of December in the location, calculated according to (A. Duffie and A. Beckman, 2013) and taking into account that the slope angle β is equal to the location latitude.

Once the distance between panels is determined, the next step is the calculation of the solar panels area. The relation between the tilt-mounted solar panels area and the sum of the horizontal projection of the panel's surface and the area corresponding with the distance between rows is named the covering ratio (CR hereinafter), which, as illustrated in the Figure 6, can be calculated as:

$$CR = \frac{w}{a+d} \qquad [1]$$

The Table 4 shows the results obtained for the covering ratio for different latitude angles between 0° y 50°. As can be seen, on the equator the optimum tilt angle is 0° and the covering ratio is one, therefore all the PV roof space can be covered with horizontal-mounted solar panels. The ratio decreases while the latitude increases reaching the minimum for a latitude of 50° where only 0.2 m^2 of solar panels can be installed per every square meter of PV roof space.

An average covering ratio is applied to the PV roof space according to the location to obtain the total solar panels area, called PV Area, for every single location included in the study. The results are shown in the Table 5.

The peak power installable in the roof PV facility (P_P) will be determined by the characteristics of the solar panel, that is:

Table 5. Calculation of the PV area

Region	PV Roof space	Latitude (φ) range	Average Covering ratio	PV area (A_{PV})
	(Ha)	(deg)		(Ha)
US	17,028	30°-45° N	0.45	7,663
EU28	4,290	35°-60° N	0.36	1,544
China	1,481	15°-45° N	0.76	1,126
Australia	544	15°-45° S	0.76	413
India	181	0°-30° N	0.78	141
Total	**23,524**			**10,887**

$$P_p = A_{PV} \cdot \frac{P_{SP}}{A_{SP}} \qquad [2]$$

Where A_{SP} is the solar panel area and P_{SP} the maximum output power under standard conditions. Nowadays there are different technologies for the manufacturing of solar panels, the most widely used is made of multi-crystalline silicon cells (International Technology Roadmap for Photovoltaic, 2015; REN21, 2013). These solar panels have recently reached efficiency levels over 20% in laboratory (National Center for Photovoltaics, 2015). For the calculation of the P_P, we have selected a solar panel manufactured with multi-crystalline silicon cells and a certified efficiency of 15.5%. Other characteristics are shown in the Table 6.

Therefore, for the solar panel selected:

$$\frac{P_{panel}}{A_{panel}} = \frac{300\ W}{(1.956 \cdot 0.992)\ m^2} = 1.55\ \text{MW/Ha} \qquad [3]$$

La Table 7 illustrates the calculation of the power capacity. A potential of 16,833 GW is obtained, that is of the same order of magnitude as the total capacity installed in US ending 2014. Taking into consideration that the total power installed at the end of 2104 in the countries included in this article was 164,272 GW, the shopping centers PV potential means 10%.

Table 6. Solar panel characteristics

Characteristic	Value
Manufacturer	Trina Solar
Model	TSM-PC14
Cell type	Si Multicrystalline
Max Power (STC conditions)	300 W
Efficiency (ε)	15,5 %
Dimensions (h x w x d)	1,956 x 992 x 40 mm^3

Table 7. PV power potential in shopping centers

Region	Solar panels area (A_{PV}) (Ha)		$\dfrac{P_{SP}}{A_{SP}}$		SC potential (GW)	Present installed (GW)	SC potential vs 2014
US	7,663				11,848	18,317	65%
EU28	1,544				2,388	86,796	3%
China	1,126	x	1.55	=	1,740	28,050	6%
Australia	413				639	28,050	2%
India	141				218	3,059	7%
Total	**10,887**				**16,833**	**164,272**	**10%**

The yearly electricity generation in the shopping centers can be estimated using the following expression:

$$E = G_T \cdot A_{PV} \cdot PR \cdot \varepsilon \qquad [4]$$

Where G_T is the total yearly solar irradiation incident on an optimally-tilted solar panel (in this study equal to the site latitude), ε is the solar panel efficiency and PR is the facility performance ratio.

There are many sources to obtain the global radiation in a flat surface tilted at latitude angle. In this study, the data used in the calculations have been extracted from (Australian and New Zealand Solar Energy Society, 2006; Huld et al., 2012; National Renewable Energy Laboratory, 2015), then an average global radiation was estimate for every single country or territory.

The PV facilities present PR values in the 60-90% range (Dierauf et al., 2013; Reich et al., 2012). In this study, an average value of 75% was considered for PR. Finally, for the solar panel efficiency, the value considered is the one included in the Table 6.

Other factors might be considered in the calculation of the energy generation. One of them is the degradation of the solar panels, effect that could be limited with an appropriate maintenance of the facilities, including when it is necessary, the replacement of affected units. Other is the dependency of the generation with the environmental conditions; especially the temperature, that affects to the output power of the solar panels in an inversely proportional way.

In this matter, due to the general perspective of this research and the wide area covered, it would not be recommendable to set up here specific environmental conditions to be applied to the calculations because they could affect diversely (increasing the generation in some places and decreasing it in others); so that, the real effect should be evaluated for every single facility independently.

The Table 8 shows the results obtained for the electricity generation calculated with the equation [5]. Based on the hypothesis set up in this study, it would be possible a yearly electricity production of 22,783 GWh, that is 0.16% of total electricity consumption in 2014 in the territorial range of this article.

SHOPPING CENTERS ELECTRICITY CONSUMPTION

According to CBECS (U.S. Energy Information Administration, 2003), the yearly average electricity consumption per square meter of floor-space[3] in shopping centers larger than 50,000 sq ft (4,645 m^2) is 41,5 kWh/m^2. As estimated in (International Council of Shopping Centers), the relation between GLA and floor-space is in the 0.5-0.65 range. In this study it has been considered an average value of 0.6.

The Table 9 shows the estimate of the yearly electricity consumption in the shopping centers located in the countries included in the study and the share that could be supply by the PV facility. The results indicate that it would be possible to cover around 7% of the total electricity demand, with special relevance in Australia and India, where the PV share might reach 11%.

UNCERTAINTY ABOUT THE RESULTS

In the methodology proposed in this chapter, three coefficients have been used to reduce the total GLA space available in shopping centers to obtain the PV roof space: the suitability ratio, the roof space/GLA ratio and the availability ratio.

---

[3] In CBECS the floor-space is defined as all the area enclosed by the exterior walls of a building, both finished and unfinished, including indoor parking facilities, basements, hallways, lobbies, stairways, and elevator shafts

Among these three coefficients, the roof space/GLA ratio was set up by means of the analysis of a wide sample of 500 shopping centers carried out specifically by the authors for this research, and the availability ratio was determined from reliable research carried out for others and widely used in the late technical literature.

As a result of the aforementioned, the major degree of uncertainty over the results obtained in this chapter arises from the determination of the suitability ratio. The application of the methodology shows that the final results obtained in chapter 0 for PV power capacity and generation potential are linear with the value considered for the suitability ratio. Conscious of the impact of this ratio in the results, it was set up with a conservative value of 0.8

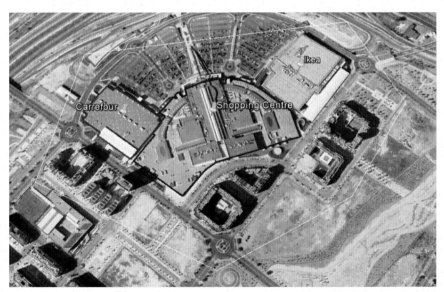

Source: Google earth ™.

Figure 7. Shopping Centre La Gavia, IKEA and Carrefour centers (Madrid, Spain).

Table 8. PV generation potential in shopping centers

Region	Solar panels area (A_{PV}) (Ha)	Average irradiance (G_T) (kWh/m²/year)	SC potential (GWh/year)	Yearly PV Generation	
				Total in 2014 (GWh/year)	SC potential vs Total in 2014
US	7,663	2,008	17,306	4,297,300	0.40%
EU28	1,544	1,300	2,259	3,108,600	0.07%
China	1,126	1,643	2,080	5,649,600	0.04%
Australia	413	1,825	848	244,500	0.35%
India	141	1,825	290	1,208,400	0.02%
Total	10,887		22,783	14,508,400	0.16%

Table 9. Consumption and PV coverage calculation

Region	GLA (Ha)		Floor-space/GLA		Floor-space (Ha)		SC consumption intensity (GWh/Ha)		SC consumption (GWh/year)	SC potential (GWh/year)	PV share
US	61,166				101,943				244,128	17,306	7%
EU28	15,410	×	0.6	=	25,684	×	2.395	=	61,506	2,259	4%
China	5,320				8,867				1,234	2,080	10%
Australia	1,953				3,255				7,795	848	11%
India	650				1,083				2,594	290	11%
Total	84,499				140,831				337,257	22,783	7%

Source: (International Council of Shopping Centers; U.S. Energy Information Administration, 2003) and self-elaboration.

AN EXAMPLE OF APPLICATION.
LA GAVIA SHOPPING CENTRE

The research carried out in this chapter has included shopping centers located in many countries and different locations, therefore it presents a wide and diversified specter of application. However, we have considered illustrative applying the methodology with an example that represents the key advantages of these type of commercial buildings for the installation of PV facilities. The shopping center selected is named La Gavia (Klepierre Management, 2015). It is located in a high populated neighborhood of Madrid (Spain), far from the nearest electricity generation plants. Enclosed in the same outdoor area there are two accessories buildings: a Carrefour store (Centros Comerciales Carrefour, 2015) and an Ikea store (Inter IKEA Systems B.V., 2015) centers. The Figure 7 shows a Google Earth[TM] (Google Inc., 2013) satellite image of La Gavia shopping center (bordered in blue) and the two accessory buildings (bordered in red and green). In total, the complex has over 77,000 m^2 of roof space.

The application of our methodology to the Gavia shopping center gives a PV capacity in the complex of 2.94 MW able to produce 4.291 MWh per year; what means 11% of the estimate shopping center yearly consumption. The calculation details are included in the Table 10.

CONCLUSION

The penetration of PV is still far from the targets set up for this technology to contribute to the mitigation of GHG emissions responsible of climate change.

The installation of PV facilities in building offers the possibility of producing electricity near the consumption point. In terms of building

**Table 10. Application of the methodology to La Gavia shopping center
and the accessory buildings**

Concept	Units	Ikea	Carrefour	Shopping center
Roof space	(m²)	16,922	13,470	46,990
Availability ratio		0.6		
PV Roof space	(m²)	10,153	8,082	28,194
Latitude (φ)	(deg)	40° 22'		
Optimum slope angle	(deg)	35°		
Average covering ratio		0.41		
PV area (A$_{PV}$)	(m²)	4,163	3,314	11,560
Power potential	(MW)	0.64	0.51	1.79
Total power potential	(MW)	**2.94**		
Average irradiance (GT)	(kWh/m²/year)	2,029		
Generation potential	(MWh/year)	982	782	2,727
Total generation potential	(MWh/year)	**4,491**		
Floor-space	(m²)	174,319		
Consumption estimate	(MWh/year)	41,745		
PV Share		**11%**		

integration, the shopping centers are one the best choices to allocate PV because their typology and constructive characteristics may allow that, under similar conditions, the installable power capacity were triple than in residential buildings. Nevertheless, in order to estimate their own PV potential, other type of buildings, as schools, hospitals and industrial buildings, should be target of further research.

PV facilities integrated in shopping centers could contribute largely to achieve the targets, because the 16,800 GW of power capacity potential obtained is equivalent to 10% of the total PV capacity installed at the end of 2014. In the US, where presently there is only around 325 MW installed in commercial buildings, it could be possible the installation of around 11.8 GW, that is 65% of total power capacity installed in the whole country ending 2014. The PV facilities installable in the territorial scope of this study would be able to produce 22.7 TWh yearly, meaning 0.16% to the total electricity demand or 14% of the PV share in 2014.

If the PV roof facilities were used for self-consumption, they could cover about 7% on average of electricity demand, although in some countries the PV share could reach 11%.

The shopping centers are also an efficient choice if the surplus electricity is fed to the grid. They are typically located near large populated areas with high energy demand, and the generation could reach the consumption points in an efficiently way with low losses.

The installation of the PV potential obtained in this study on shopping center roofs could avoid the use of over 23,000 Ha of land, a surface larger than the area covered for the cities of New York and Washington together.

REFERENCES

[1]   A. Duffie, J. & A. Beckman, W. (2013). *Solar Engineering of Thermal Processes*, 4 ed. John Wiley & Sons, Inc., Hoboken, New Jersey.

[2]   Arbeitsgemeinschaft Energiebilanzen, (2014). *Stromerzeugung nach Energieträgern 1990 – 2014 [Power generation by energy sources 1990 - 2014]* <https://www.ag-energiebilanzen.de/4-1-Home.html>.

[3]   Australian and New Zealand Solar Energy Society. (2006). *Australian solar radiation data handbook.*

[4]   Australian Energy Regulator. (2014). *State of the energy market 2014.*

[5]   Barth, B., Concas, G., Binda Zane, E., Franz, O., Frías, P., Hermes, R., Lama, R., Loew, H., Mateo, C., Rekinger, M., Michele Sonvilla, P. & Vandenbergh, M. (2014). *PV GRID. Final project report.* August 2014. <www.pvgrid.eu>.

[6]   British Petroleum. (2015). *BP Statistical Review of World Energy.* June 2015. <www.bp.com/statisticalreview>.

[7]   Byrne, J., Taminiau, J., Kurdgelashvili, L. & Kim, K. N. (2015). A review of the solar city concept and methods to assess rooftop solar electric potential, with an illustrative application to the city of Seoul. *Renewable and Sustainable Energy Reviews*, *41*, 830-844. http://dx.doi.org/10.1016/j.rser.2014.08.023

[8]   Centros Comerciales Carrefour, S. A. (2015). Carrefour La Gavia [Online]. Available: http://www.carrefour.es/tiendas-carrefour/hipermercados/carrefour/la_gavia.aspx#cerrar [Accessed 2015].

[9]   Colmenar-Santos, A., Campíñez-Romero, S., Pérez-Molina, C. & Castro-Gil, M. (2012). Profitability analysis of grid-connected photovoltaic facilities for household electricity self-sufficiency. *Energy Policy*, *51*, 749-764. http://dx.doi.org/10.1016/ j.enpol. 2012.09.023.

[10] Cushman & Wakefield. (2014). Global shopping center development report. Spring 2014. <www.cushmanwakefield. com>.

[11] Defaix, P. R., van Sark, W. G. J. H. M., Worrell, E. & de Visser, E. (2012). Technical potential for photovoltaics on buildings in the EU-27. *Solar Energy*, *86*, 2644-2653. http://dx.doi.org/ 10.1016/j.solener.2012.06.007.

[12] Denholm, P. M., Robert, (2008). Supply Curves for Rooftop Solar PV-Generated Electricity for the United States. National Renewable Energy Laboratory, *Report No.: NREL/TP-6A0-44073*.

[13] Dierauf, T., Growitz, A., Kurtz, S., Becerra Cruz, J. L., Riley, E. & Hansen, C. (2013). Weather-Corrected Performance Ratio National Renewable Energy Laboratory, SS13.4510, T.N.; April 2013. *Report No.: NREL/TP-5200-57991.* <https:// www.nrel. gov/docs/fy13osti/57991.pdf>.

[14] DTZ. (2013). *European Retail Guide Shopping Centres.* <http://www.dtz.dk/files/other/Markedsrapport/dtz_european_reta il_guide_-_shopping_centres_march_2013.pdf>.

[15] European Climate Foundation. (2010). *Roadmap 2050: a practical guide to a prosperous, low carbon Europe.* <http://www.roadmap2050.eu/project/roadmap-2050>.

[16] Feldman, D., Boff, D. & Margolis, R. (2015). *National Survey Report of PV Power Applications in the United States 2014.* Paris: International Energy Agency.

[17] Feldman, D., Margolis, R. & Boff, D. (2014). Q2/Q3 *'14 Solar Industry Update.* U.S. Department of Energy. <energy.gov/sunshot>.

[18] Google Inc. (2013). *Google Earth* (Version 7.1.2.2041).

[19] Gunerhan, H. & Hepbasli, A. (2007). Determination of the optimum tilt angle of solar collectors for building applications. *Building and Environment*, *42*, 779-783. http://dx.doi.org/10.1016/j.buildenv.2005.09.012.

[20] Gutschner, M., Nowak, S., Ruoss, D., Toggweiler, P. & Schoen, T. (2002). *Potential for building integrated photovoltaics.* International Energy Agency, Photovoltaic Power Systems Programme. Task 7; Report No.: PVPS T7-4.

[21] Hay, J. E. & Davies, J. A. (1980). Calculation of the Solar Radiation Incident on an Inclined Surface, in *Proceedings of the First Canadian Solar Radiation Data Workshop*, Toronto, Canada, 1980, p. 59.

[22] Hernández-Moro, J. & Martínez-Duart, J. M. (2013). Analytical model for solar PV and CSP electricity costs: Present LCOE values and their future evolution. *Renewable and Sustainable Energy Reviews*, *20*, 119-132. http://dx.doi.org/10.1016/j.rser.2012.11.082.

[23] Hottel, H. C. & Woertz, B. B. (1942). Performance of Flat-Plate Solar Heat Collectors. *Transactions of the ASME*, *64*, 64-91.

[24] Huld, T., Müller, R. & Gambardella, A. (2012). A new solar radiation database for estimating PV performance in Europe and Africa. *Solar Energy*, *86*, 1803-1815. http://dx.doi.org/10.1016/j.solener.2012.03.006.

[25] Inter IKEA Systems B. V. (2015). IKEA Ensanche de Vallecas [Online]. *IKEA.* Available: http://www.ikea.com/es/es/store/ ensanche_de_vallecas/indexPage [Accessed 4/10/2015].

[26] Intergovernmental Panel on Climate Change. (2011). *Special Report on Renewable Energy Sources and Climate Change Mitigation. United Kingdom and New York, NY, USA:* Cambridge University Press.

[27] Intergovernmental Panel on Climate Change. (2014). Climate Change 2014: Synthesis Report. *Contribution of Working Groups I, II and III to the Fifth Assessment Report of the Intergovernmental Panel on Climate Change.* Geneva, Switzerland. <http://www.ipcc.ch/report/ar5/syr/>.

[28] *International Council of Shopping Centers, Asia-Pacific Shopping Centre Classifications.* New York.

[29] International Council of Shopping Centers. (2015a). *The socio-economic contribution of European shopping centers.* < https:// www.icsc.org/uploads/research/general/European-Impact-Study-2015.pdf?utm_source=research-homepage&utm_medium= web&utm_campaign=European-Impact-Study-2015>.

[30] International Council of Shopping Centers. (2015b). *U.S. Shopping-Center Classification and Characteristics.* < https:// www.icsc.org/uploads/research/general/US_CENTER_CLASSIFI CATION.pdf>.

[31] International Energy Agency. (2014a). *IEA Energy Technology Perspectives 2014.* Paris, France. <http://www.iea.org/etp/ etp2014/>.

[32] International Energy Agency. (2014b). *Technology Roadmap Solar Photovoltaic Energy.* 2014 Edition. Paris, France.

[33] International Renewable Energy Agency. (2015). *Resource.* http://resourceirena.irena.org/gateway/.

[34] International Technology Roadmap for Photovoltaic. (2015). *ITRPV 2014 Results.* April 2015.

[35] Izquierdo, S., Rodrigues, M. & Fueyo, N. (2008). A method for estimating the geographical distribution of the available roof surface area for large-scale photovoltaic energy-potential evaluations. *Solar Energy*, *82*, 929-939. http://dx.doi.org/ 10.1016/j.solener.2008.03.007.

[36] Johnston, W., Taeni, C. & Egan, R. (2015). National Survey Report of PV Power Applications in Australia 2014. Paris: International Energy Agency.

[37] Karteris, M., Slini, T. & Papadopoulos, A. M. (2013). Urban solar energy potential in Greece: A statistical calculation model of suitable built roof areas for photovoltaics. *Energy and Buildings*, *62*, 459-468. http://dx.doi.org/10.1016/j.enbuild.2013.03.033.

[38] Klepierre Management. (2015). Shopping Center La Gavia [Online]. Available: http://es.club-onlyou.com/La-Gavia [Accessed 2015].

[39] Klucher, T. M. (1979). Evaluating Models to Predict Insolation on Tilted Surfaces. *Solar Energy*, *23*.

[40] Lazard. (2014). *Levelized Cost of Energy* v8. <https:// www.lazard.com/perspective/levelized-cost-of-energy-v8-abstract/>.

[41] Lewis, G. (1987). Optimum tilt of a solar collector. *Solar & Wind Technology*, *4*, 407-410. http://dx.doi.org/10.1016/0741-983X (87)90073-7.

[42] Li, D., Liu, G. & Liao, S. (2015). Solar potential in urban residential buildings. *Solar Energy*, *111*, 225-235. http://dx. doi.org/10.1016/j.solener.2014.10.045.

[43] Liu Y. H., B., C. & Jordan, R. (1960). The Interrelationship and Characteristic Distribution of Direct, Diffuse and Total Solar Radiation. *Solar Energy*, *4*.

[44] Luthander, R., Widén, J., Nilsson, D. & Palm, J. (2015). Photovoltaic self-consumption in buildings: A review. *Appl. Energy*, *142*, 80-94. http://dx.doi.org/10.1016/ j.apenergy. 2014.12.028.

[45] Mai, T., Wiser, R., Sandor, D., Brinkman, G., Heath, G., Denholm, P., Hostick, D. J., Darghouth, N., Schlosser, A. & Strzepek, K. (2012). *Exploration of High-Penetration Renewable Electricity Futures*. Golden, CO: Report No.: NREL/TP-6A20-52409-1. <http://www.nrel.gov/docs/fy12osti/52409-2.pdf>.

[46] National Center for Photovoltaics. (2015). Research Cell Efficiency Records. *U.S. Department of Energy*. <http://www.nrel.gov/ncpv/>.

[47] National Renewable Energy Laboratory. (2012). SunShot Vision Study. U.*S. Department of Energy, Report No.: DOE/GO-102012-3037*. <www.energy.gov/sunshot>.

[48] National Renewable Energy Laboratory. (2015). *MapSearch*. Retrieved 2015 http://www.nrel.gov/gis/mapsearch/.

[49] Ondraczek, J., Komendantova, N. & Patt, A. (2015). WACC the dog: The effect of financing costs on the levelized cost of solar PV power. *Renewable Energy*, *75*, 888-898. http://dx.doi.org/10.1016/j.renene.2014.10.053.

[50] Ordóñez, J., Jadraque, E., Alegre, J. & Martínez, G. (2010). Analysis of the photovoltaic solar energy capacity of residential rooftops in Andalusia (Spain). *Renewable and Sustainable Energy Reviews*, *14*, 2122-2130. http://dx.doi.org/10.1016/j.rser.2010.01.001.

[51] Ossenbrink, H., Huld, T., Jäger Waldau, A. & Taylor, N. (2013). Photovoltaic Electricity Cost Maps. *Report No.: JRC 83366*.

[52] Peng, J. & Lu, L. (2013). Investigation on the development potential of rooftop PV system in Hong Kong and its environmental benefits. *Renewable and Sustainable Energy Reviews*, *27*, 149-162. http://dx.doi.org/10.1016/j.rser.2013.06.030.

[53] Perez, R., Ineichen, P., Seals, R., Michalsky, J. & Stewart, R. (1990). Modeling daylight availability and irradiance components from direct and global irradiance. *Solar Energy*, *44*, 271-289. http://dx.doi.org/10.1016/0038-092X(90)90055-H.

[54] Perez, R., Seals, R., Ineichen, P., Stewart, R. & Menicucci, D. (1987). A New Simplified Version of the Perez Diffuse Irradiance Model for Tilted Surfaces. *Solar Energy, 39*.

[55] Perez, R., Stewart, R., Seals, R. & Guertin, T. (1988). The Development and Verification of the Perez Diffuse Radiation Model. Oct-1988. *Report No.: SAND88-7030.*

[56] Philipps, S. P., Kost, C. & Schlegl, S. (2014). *Up-to-date levelised cost of electricity of photovoltaics.* Fraunhofer Institute for Solar Energy Systems ISE.

[57] Pillai, I. R. & Banerjee, R. (2007). Methodology for estimation of potential for solar water heating in a target area. *Solar Energy, 81,* 162-172. http://dx.doi.org/10.1016/j.solener.2006.04.009.

[58] Red Eléctrica de España, S. A. (2015). Informe del Sistema Eléctrico Español 2014. [Spanish Electric System Report 2014.] Madrid. <http://www.ree.es/es/ publicaciones/sistema-electrico-espanol/informe-anual/informe-del-sistema-electrico-espanol-2014#>.

[59] RegioData Research. (2013). EU Shopping centers factsheets. *RegioData Research.* <http://www.retailcenters.eu/factsheets>.

[60] Reich, N. H., Mueller, B., Armbruster, A., van Sark, W. G. J. H. M., Kiefer, K. & Reise, C. (2012). Performance ratio revisited: is PR > 90% realistic? *Progress in Photovoltaics: Research and Applications, 20,* 717-726. http://dx.doi.org/10.1002/pip.1219.

[61] Reindl, D. T., Beckman, W. A. & Duffie, J. A. D. (1990). Evaluation of Hourly Tilted Surface Radiation Models. *Solar Energy, 45.*

[62] REN21. (2013). *Renewables Gobal Future Report.* Paris. <http://www.ren21.net/future-of-renewables/global-futures-report/>.

[63] REN21. (2015). *Renewables 2015 Global Status Report.* Paris. <http://www.ren21.net/status-of-renewables/global-status-report/>.

[64] RTE Réseau de transport d'électricité. (2015). 2014 Annual Electricity Report. <http://www.rte-france.com/en/news/elect ricity-report-2014-drop-electricity-demand-and-increase-renew able-energies-which>.

[65] Schallenberg-Rodriguez, J. (2013). Photovoltaic techno-economical potential on roofs in regions and islands: The case of the Canary Islands. Methodological review and methodology proposal. *Renewable and Sustainable Energy Reviews*, *20*, 219-239. http://dx.doi.org/10.1016/j.rser.2012.11.078.

[66] Singh, R. & Banerjee, R. (2015). Estimation of rooftop solar photovoltaic potential of a city. *Solar Energy*, *115*, 589-602. http://dx.doi.org/10.1016/j.solener.2015.03.016.

[67] Solar Energy Industries Association. (2014). *Solar Means Business 2014*. Top U.S. Commercial Solar Users. Washington DC.

[68] Stetz, T., Rekinger, M. & Theologitis, I. (2014). Transition from uni directional to bi directional distribution grids. Management summary of IEA Task 14 Subtask 2 - Recommendations based on *Global Experience*. September, 2014. *Report No.: IEA PVPS Task 14:03-2014.*

[69] The US Department of Energy. (2015). *The US Department of Energy* [Online]. Available: https://www.energy.gov/ [Accessed 2015].

[70] The White House. President of the US. (2013). *The President's Climate Action Plan.* Executive Office of the President.

[71] U.S. Energy Information Administration. (2003). Commercial Buildings Energy Consumption Survey. *U.S. Energy Information Administration.*

[72] U.S. Energy Information Administration. (2015). *Electricity data browser* [Online]. Available: https://www.eia.gov/electricity/ data/browser/ [Accessed 14/07/2015 2015].

[73] U.S. Environmental Protection Agency. (2015). *Carbon Pollution Emission Guidelines for Existing Stationary Sources: Electric*

Utility Generating Units; Final Rule. U.S. Federal Register, pp. 64661–65120.

[74] UK Government. *Department of Energy & Climate Change.* (2015). Renewable energy in 2014. <https://www.gov.uk/government/statistics/energy-trends-june-2015-special-feature-articles-renewable-energy-in-2014>.

[75] Wiginton, L. K., Nguyen, H. T. & Pearce, J. M. (2010). Quantifying rooftop solar photovoltaic potential for regional renewable energy policy. *Computers, Environment and Urban Systems*, *34*, 345-357. http://dx.doi.org/10.1016/j.compenvurbsys.2010.01.001.

[76] Yamada, H. & Ikki, O. (2015). *National Survey Report of PV Power Applications in Japan 2014*. Paris.

ABOUT THE EDITORS

Antonio Colmenar Santos, PhD
Full Professor
Departamento de Ingeniería Eléctrica, Electrónica, Control,
Telemática y Química Aplicada
Universidad Nacional de Educación a Distancia, Madrid, Spain
Email: acolmenar@ieec.uned.es

Dr. Colmenar-Santos has been a senior lecturer in the field of Electrical Engineering at the Department of Electrical, Electronic and Control Engineering at the National Distance Education University (UNED) since June 2014. Dr. Colmenar-Santos was an adjunct lecturer at both the Department of Electronic Technology at the University of Alcalá and at the Department of Electric, Electronic and Control Engineering at UNED. He has also worked as a consultant for the INTECNA project (Nicaragua). He has been part of the Spanish section of the International Solar Energy Society (ISES) and of the Association for the Advancement of Computing in Education (AACE), working in a number of projects related to renewable energies and multimedia systems applied to teaching. He was the coordinator of both the virtualisation and telematic Services at ETSII-UNED, and deputy head teacher and the head of the Department of Electrical, Electronics and Control Engineering at UNED. He is the author of more than 50 papers published in respected journals (http://goo.gl/ YqvYLk) and has participated in more than 100 national and international conferences.

David Borge Diez, PhD
Lecturer
Departamento de Ingeniería Eléctrica y de Sistemas y Automática
Universidad de León, León, Spain
Email: dbord@unileon.es

Dr. David Borge-Diez has a PhD in Industrial Engineering and a MSc. in Industrial Engineering, both from the School of Industrial Engineering at the National Distance Education University (UNED). He is currently a lecturer and researcher at the Department of Electrical, Systems and Control Engineering at the University of León, Spain. He has been involved in many national and international research projects investigating energy efficiency and renewable energies. He has also worked in Spanish and international engineering companies in the field of energy efficiency and renewable energy for over eight years. He has authored more than 25 publications in international peer-reviewed research journals and participated in numerous international conferences.

Enrique Rosales Asensio, PhD

Researcher
Departamento de Física
Universidad de La Laguna, La Laguna, Spain
Email: erosalea@ull.edu.es

Dr. Rosales-Asensio, PhD, is an industrial engineer with postgraduate degrees in electrical engineering, business administration, and quality, health, safety and environment management systems. Currently, he is a senior researcher at the University of La Laguna, where he is involved in water desalination project in which the resulting surplus electricity and water would be sold. He has also worked as a plant engineer for a company that focuses on the design, development and manufacture of waste-heat-recovery technology for large reciprocating engines; and as a project manager in a world-leading research centre.

INDEX

Design, Modeling, Manufacturing and Performance Evaluation of a Solar-Powered Single-Effect Absorption Cooling SystemSignificance

Author: Vahid Vakiloroaya (School of Computing, Engineering and Mathematics, Western Sydney University, Australia)

Series: Energy Science, Engineering and Technology

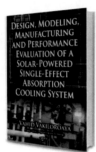

Book Description: Conventional HVAC systems rely heavily on energy generated from fossil fuels, which are being rapidly depleted. This – together with a growing demand for cost-effective infrastructure and appliances – has necessitated new installations and major retrofits in occupied buildings to achieve energy efficiency and environmental sustainability.

Hardcover ISBN: 978-1-53610-882-8
Retail Price: $160

Space-Based Solar Power: Feasible Idea or Folly?

Editor: Carl P. Thompson

Series: Energy Science, Engineering and Technology

Book Description: This book examines the current progress of space-based power in the areas of technology, economics, and operations, then contemplates its role in the U.S. grand strategy for space.

Softcover ISBN: 978-1-63483-145-1
Retail Price: $69

SOLAR POWER: TECHNOLOGIES, ENVIRONMENTAL IMPACTS AND FUTURE PROSPECTS

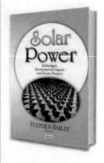

EDITOR: Stephen Bailey

SERIES: Energy Science, Engineering and Technology

BOOK DESCRIPTION: This book examines the renewable energy policies in Germany, the Netherlands, Spain, Switzerland and the UK in order to identify the policy motivations and constraints for expanding renewables in these countries.

HARDCOVER ISBN: 978-1-63321-317-3
RETAIL PRICE: $230

SHARED SOLAR POWER: PROJECT GUIDES FOR COMMUNITIES

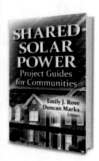

EDITORS: Emily J. Rose and Duncan Marks

SERIES: Energy Science, Engineering and Technology

BOOK DESCRIPTION: "Community shared solar" is defined as a solar-electric system that provides power and/or financial benefit to multiple community members. This book provides current detail on projects designed to increase access to solar energy and to reduce up-front costs for participants.

HARDCOVER ISBN: 978-1-62257-527-5
RETAIL PRICE: $130